Methodology for the Development of Sustainable Products

Roland Lachmayer · Johanna Wurst ·
Jorin Thelemann

Methodology for the Development of Sustainable Products

Springer

Roland Lachmayer
Institut für Produktentwicklung und Gerätebau (IPeG)
Leibniz Universität Hannover
Garbsen, Germany

Johanna Wurst
Institut für Produktentwicklung und Gerätebau (IPeG)
Leibniz Universität Hannover
Garbsen, Germany

Jorin Thelemann
Institut für Produktentwicklung und Gerätebau (IPeG)
Leibniz Universität Hannover
Garbsen, Germany

ISBN 978-3-662-72441-5 ISBN 978-3-662-72442-2 (eBook)
https://doi.org/10.1007/978-3-662-72442-2

This book is a translation of the original German edition "Entwicklungsmethodik nachhaltiger Produkte" by Roland Lachmayer et al., published by Springer-Verlag GmbH, DE in 2025. The translation was done with the help of an artificial intelligence machine translation tool. A subsequent human revision was done primarily in terms of content, so that the book will read stylistically differently from a conventional translation. Springer Nature works continuously to further the development of tools for the production of books and on the related technologies to support the authors.

Responsible Editor: Alexander Grün

This Springer imprint is published by the registered company Springer-Verlag GmbH, DE, part of Springer Nature.
The registered company address is: Heidelberger Platz 3, 14197 Berlin, Germany

Preface

Many current politically influenced discussions about the necessity and consequences of technical products for our society and the planet Earth, and the often extremely visionary hopes for sustainable solutions to needs through technical innovations, are reason enough for this book. However, the decisive factor in starting work on this book was the decision of the Faculty of Mechanical Engineering at Leibniz University Hannover to offer the course "Sustainable Engineering" from the winter semester 2021, along with the challenge presented to us in this context to deliver a lecture on "Sustainable Product Engineering—Development of Sustainable Products and Student Design Project".

To avoid misunderstandings, it should be explained that this book is about the technical development of sustainable products—in English "Sustainable Engineering Design" or "EcoDesign"—not about design in the sense of form design, and that the sustainable working environment in a development department, e.g., the concept of the paperless office, is only touched upon marginally.

When considering the development methodology for sustainable products, it quickly becomes clear that the current discussion, which is strongly focused on global warming, falls far short of the actual complexity of the topic. There are numerous findings on dealing with raw materials, toxins, waste, and burdens on flora and fauna worldwide in almost countless laws, standards, and rules. Many established test and evaluation procedures need to be considered in general and in particular. The book will also discuss these, as well as the questions: What characterises sustainable products, apart from compliance with legal regulations, and how can objective evaluations of these be carried out?

Our book is not only a textbook but also a documentation of our scientific work. We are particularly interested in: How do we arrive at solutions, and which aspects of the development methodology are particularly relevant for the objective of developing sustainable products at the levels of specification, concept development, design, and detail design, or still need to be researched?

Finally, the book includes examples and lessons learned, many of which are drawn from our industry and student projects, intended to encourage questioning of existing solutions.

My special thanks goes to all those who have supported us intellectually and editorially, thereby contributing to the success of this book.

Hannover, Germany
September 2025

Univ.-Prof. Dr.-Ing. Roland Lachmayer

Competing Interests The authors have no competing interests to declare that are relevant to the content of this manuscript.

Contents

About the Authors

Univ.-Prof. Dr.-Ing. Roland Lachmayer has been and continues to be active as Director of the Institute for Product Development and Device Construction (IPeG) since 2010 and Senator at Leibniz University Hannover since 2020. He is Managing Director of the Scientific Society for Product Development, a member of the Advisory Board "Future of Value Creation" of the BMFTR (Federal Ministry of Research, Technology and Space), a member of the Scientific Directorate of the Laser Centre Hannover e. V., and a board member for the research building Scale (Scalability of Production Processes). Roland Lachmayer has been and continues to be actively involved in numerous funded research projects, including SFBs 653 and 1153, the European-funded research consortia GROTESK and AM2H2, the graduate schools Tailored Light and SAM, and the Excellence Cluster PhoenixD. In addition to his scientific work, technology transfer through Innovation Cells, company foundations, and acquisitions has significantly shaped Prof. Lachmayer's career. His over 400 publications and over 30 patents address the research focuses of the IPeG: Modelling and Simulation in Development, Development Methodology for Additive Manufacturing, Development of Optical Devices, and Data-Driven Development. The development of sustainable products is a cross-cutting issue at the institute.

Johanna Wurst is Head of the Research Group "Data Driven Design" at the Institute for Product Development at Leibniz University Hannover and has focused the research activities on the development methodology for sustainable products in recent years. Key projects under her responsibility were the establishment of the course "Sustainable Product Design—Development of Sustainable Products" described in this book, as well as other activities to integrate a holistic understanding of sustainability into science and teaching. In her own research, she addresses the integration of ecological and economic evaluations to increase eco-efficiency in organisations and their processes.

Jorin Thelemann is a research associate at the Institute for Product Development at Leibniz University Hannover. Current research activities include the development methodology of sustainable products, EcoDesign, and the design and usage of sustainability-oriented DfR strategies. He is also involved in teaching as a project coordinator in the interdisciplinary teaching and learning project *novaIMPULS*.

Introduction 1

In a time when global resource consumption is continuously increasing and environmental problems are becoming more pressing, the topic of sustainability is increasingly coming into focus for the economy and society. Terms such as "sustainability" and "efficiency" are ubiquitous and have become buzzwords in many respects, stimulating not only discussions but also requiring concrete actions. Whether it's new health management systems, modern production lines, or the expansion of solar plants, sustainable developments are indispensable in our daily lives. Technical products, whether in the form of goods or services, are at the centre of the discussion. They are essential for progress, but they are not a perpetual motion machine—they also require resources for their production and usage. The challenge of our time is to develop engineering and mechanical solutions that are both ecologically and socially sustainable. This task is not new, but the societal and global prioritisation of environmental and social aspects has led to technical innovations being demanded with a significant technology pull to open up new possibilities. The last decade has shown how strong the influence of legislation and regulatory requirements has become to ensure ecological sustainability. Important laws such as the Act on Liability for Defective Products, the Electrical and Electronic Equipment Act and the Circular Economy Act, as well as international standards and guidelines such as ISO 14062 (Guidelines for the integration of environmental aspects in product development) and ISO 59004 (Circular Economy—Vocabulary, principles and guidelines for implementation) set the framework for sustainable product development. An increasing environmental awareness in society is reflected in these requirements and emphasises the significant role of product development in this context.

This book aims to systematically illuminate the development methodology for sustainable products. It combines established knowledge with insights and methods that have been researched and applied in the last decades, with a focus on

R. Lachmayer et al., *Methodology for the Development of Sustainable Products*,
https://doi.org/10.1007/978-3-662-72442-2_1

sustainability. A key hypothesis is that the development of sustainable products is not a special path or additional effort, but should represent the new basis. We invite the readers to explore the following essential aspects with us:

- **Sufficiency**: The reduction of products to the essentials within the context of systems and business models.
- **Legality**: Understanding and applying legal frameworks, standards, and incentives.
- **Efficiency**: Minimising the lifecycle cost, particularly concerning the ecological footprint.
- **Consistency**: Coherent innovations, materials, and concepts that promote sustainability in energy and the circular economy.
- **Embodiment design**: Assessment of product architecture and design from repair and recycling perspectives.
- **Lifecycle**: Ensuring function, lifespan, and value over the entire lifecycle.

The book begins with strategic considerations on systems and business models, leads through standards and norms, to concrete methods and calculation models. Under the motto "Shaping the future with knowledge", supplementary data and case studies offer deeper insights. Based on 15 years of industrial activity and almost the same amount of time in research and teaching, it is aimed at students and practitioners alike. It is intended to serve engineers, managers, and all technically interested parties as a comprehensive guide in the creative challenge of developing sustainable products and sharing responsibility for an environmentally friendly future.

1.1 Dimensions of Sustainability

The concept of sustainability is very old and originates in German forestry, where it has been applied for many decades:

> Sustainability is a principle of resource use, which aims to ensure long-term satisfaction of needs by preserving the natural regenerative capacity of the involved systems. (freely quoted from: Hans Carl von Carlowitz, 1713)

Applied in forestry, this means that no more wood may be taken from the forest than can regrow, thus establishing a generational contract under normal growth durations of commercial timber (60–120 years). Even three hundred years ago, people were consciously thinking about how to preserve their increasingly populated and used environment for future generations. It can also be assumed that similar considerations applied in older high cultures, although differently than today, not against the background of a global economy, but rather a local perspective. However, we can only achieve technical progress if we think within concrete, manageable system boundaries.

Think globally, act locally….

This was already formulated by the "hippie" movement in 1968 [1].

The sustainability debate was significantly advanced following the Brundtland Report, commissioned by the United Nations (UN) in 1987, and the statements of the "Club of Rome" in the 1990s [2].

On the one hand, the concept of sustainability is used in many ways today in light of a multitude of labels and concepts with primarily economic interests and previous unsuccessful efforts to achieve real improvements. On the other hand, a holistic view is about nothing more and nothing less than the question of how the civilisational project should be continued in the postmodern era. This includes considering how a reasonable and realistic balance between social, economic, and ecological aspects can be achieved to ensure the supply of humanity, maintain the stability of societies, and leave an intact and liveable world for future generations. The challenges of shaping a future that meets these requirements are diverse and far from exclusively technical in nature.

A significant problem is that all previous progress is based on a growth-oriented economy and continues to consume its prerequisites, in terms of a functioning world and available raw materials. For example, the electric mobility realised so far, which enters the market with sports and off-road vehicles, but does not have an innovative mobility concept. And, to mention this as well, e-scooters do not solve a real mobility problem, but merely enrich our city centres with another gadget, using valuable raw materials. In addition, almost all regions of the world continue to strive for affluent societies with established notions of quality of life, despite the effects that have already occurred.

The challenge seems to be not only the product itself but also the human being in his manifestation as an individual. While he postulates the "good" as a political social being with great reason and social correctness, he behaves individually counterproductively. It is well known that many people, as their standard of living increases, find their happiness in ever larger flats, ever larger cars, and ever further searching for holiday travel. Not to forget the energetic and material efforts for a constantly growing need for media and communication.

Even though we did not set out with this book to discuss the economic and social boundary conditions in our society and in those that follow the "western" model, these will always play a crucial role, especially in product definition. Under given boundary conditions, technical solutions can only be as ecological as the market allows—and this is determined not only by technical innovations but also by economic boundary conditions and purchasing behaviour. We have seen many positive developments in the industrial age (such as regulations for wastewater discharge into rivers, handling of asbestos, prohibition of leaded petrol, and the creation of bicycle infrastructures) that, in addition to technical solutions, the political element—whether through regulations or incentives—has a decisive influence. From a technical point of view, sustainability is primarily enforced by ecological requirements and defined by interventions in the market economy. From an ecological point of view, energy efficiency and the use of renewable resources must be

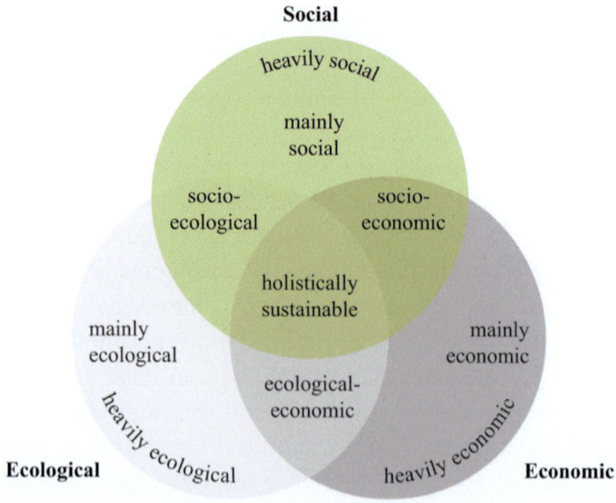

Fig. 1.1 Integrative sustainability triangle; own representation in reference to Hauff [3]

increased, and harmful materials must be avoided and kept away from our environment. Not to forget that security for people and the environment must always and primarily be taken into account. The economic approach aims to maintain general competitiveness, for example, when using innovative technologies.

The systematic description of the effects of human action from different perspectives and angles requires a classification. This classification in the form of various dimensions of sustainability and their interconnections is shown in Fig. 1.1 in reference to Hauff [3] and consists, as described, of ecological, economic, and social perspectives. Depending on context and product idea, decisions can be classified into this triangle, and thus development goals can be more clearly focused.

The integrative sustainability triangle, unlike the "Triple Bottom Line" approach, does not distinguish between three isolated dimensions of ecology, economy, and social aspects, but assumes a linkage of sustainability dimensions. This results in tension fields within the triangle that allow a focus or a more detailed definition of sustainability. For example, medical products are not only to be developed from an economic perspective. Rather, the social-societal benefit for patients influences the sustainability of the product. Specifically, a classification of the sustainability of these medical products in a socio-economic tension field can be made. At the centre of the integrative sustainability triangle is the target state of holistic sustainability. However, since individual products or organisations can neither achieve a hundred percent, holistic sustainability from a temporal nor a content-related perspective, an approximation through the usage of these tension fields is beneficial.

In the context of the sustainability discussion, the United Nations (UN), as a result of their World Summit for Sustainable Development in 2015, in their

Fig. 1.2 Sustainable development goals of the UN [4]

so-called *Sustainable Development Goals* (short: SDGs), described various objectives that are widely spread today. In the following Fig. 1.2, these 17 goals are graphically represented.

With the *Agenda 2030* adopted in 2015, the German Federal Government also commits to a concrete action plan. In the so-called *German Sustainability Strategy*, the 17 goals of the UN are taken up and transferred into a plan for realisation and goal achievement. The main concern is to develop a holistic strategy with the help of which the 17 goals can be efficiently implemented both in Germany and in other countries. The German Sustainability Strategy provides a kind of basis or framework for the transfer of the goals, for example, through political reforms, laws, and further changes. In addition, since 2009, there has been a mandatory sustainability check when new laws or regulations are drafted by all departments of the German Federal Government. As part of this check, the issuing authorities must demonstrate to what extent the effects of the legislative project are in line with sustainable development. In the course of this practice, it is also becoming increasingly common to check the respective draft laws for their compliance with the 17 global sustainability goals [5].

From the perspective of product development, SDGs, such as *Industry, Innovation and Infrastructure (9)* or *Responsible Consumption and Production Patterns (12)*, seem obvious at first glance; the combination of all SDGs forms the basis for the development of more sustainable products. Depending on the functions, applications, and resulting life cycles of the products, different and individually selectable goal combinations make sense.

In engineering ethics, similar objectives are often found, as formulated by the UN. This is done in a more or less detailed description—here is a short excerpt: “Engineers are the intellectual parents with whose help scientific knowledge is applied for the practical benefit of mankind”. It should be noted that engineers, as individuals, set their priorities first based on considerations of their person, then on their family and friends, followed by their immediate environment (e.g., company), the community, and finally the global environment.

1.2 Sustainable Products

When entering the term "sustainable product" into well-known search engines, the search initially yields glass straws, bamboo toothbrushes, cotton bread bags, or an LED light named *Little Sun*. A simple answer to the question "What actually makes a product sustainable?" is not immediately apparent.

Therefore, an attempt will be made to define this, noting that there are at least six aspects of sustainability that can be combined with the term product:

Primarily sustainable products are those that grow completely independently and based on natural resources, such as an apple or, more technically, a piece of wood. However, if fertilisers and pesticides are used, harvesting machines are used, and packaging, logistics, and distribution in a lit and heated store or cooking food are added, even these products are only conditionally sustainable in terms of their consumption. Coffee is an example that is often criticised in this context.

Cyclically sustainable products are those whose materials—such as metals or plastics—are not primarily sustainable as they are based on fossil resources. However, these materials were once in a cycle or still are and are reused through recycling strategies. This approach of cyclically sustainable products forms the basis of the strategy of the circular economy.

Relatively sustainable products are those that are more sustainable than their predecessor generation, such as a vacuum cleaner with lower material and/or energy consumption while maintaining or increasing performance.

Functionally sustainable products are those that are not particularly sustainable in their construction, but contribute to a sustainable overall system in their function. Examples include batteries for electric vehicles or the use of polysilicon for the production of solar cells.

Partially sustainable products are those in whose overall system a part was realised with particular consideration of sustainability. An example here could be coconut mats, which are used instead of plastics for sound insulation in a vehicle interior, regardless of the type of drive or the purpose of the vehicle.

Finally, there are **seemingly sustainable products**, where external impact and clever ideas are sold as presented in the examples listed at the beginning. The proof of an ecological, economic, or social effect that is comprehensible to customers is often not provided, but rather these are usually "gadgets".

Especially with more complex systems, mixed forms of the above-mentioned aspects of more sustainable products often occur. As a rule, it is often challenging to assess how good something really is in terms of sustainability without the evaluation procedures explained in more detail in Chap. 5. This is because interconnected systems, entire life cycles, and associated supply chains of products must be analysed and evaluated. In addition, when replacing an existing product,

Fig. 1.3 Survey results of a survey on the value of products (Mentimeter survey; lecture "Development of sustainable products"; 105 participants)

the new one must be significantly "better" to justify additional production effort. This improvement does not have to address the holistic concept of sustainability, but rather specifically encompasses a socio-ecological, strongly economic, or predominantly social change.

Against the background of resource consumption and environmental destruction, of course, those products are most sustainable that do not become real at all—that we do not need at all—or for other reasons do not produce at all. Indeed, the question "How necessary are all the new things on the market?" is certainly not unjustified. Does this mean in return that we have to and should go back to the Stone Age? Certainly not, because it is well known that not everything was better in the past, as the average life expectancy shows.

We want to delve deeper into this topic using a long-known model from sociology—the *Maslow's hierarchy of needs*:

The American psychologist Abraham Maslow describes in a simplified way a hierarchical structure of human needs and motivations. These needs are influenced not least by individual values and their manifestations. Based on a survey of young engineering students (anonymous, approx. 105 participants), concrete manifestations of these values are shown (see Fig. 1.3).

Distinguished by personal benefit, emotionality, monetary and social values, a broad spectrum of needs is revealed that must be addressed and satisfied by products.

The value of a product is often very technically conceived in product development and defined as the ratio of benefit to the incurred costs. A widely accepted definition comes from VDI 2800, which describes the value of a product as the "benefit per expenditure", where the benefit is determined by the fulfilment of customer requirements and the quality of the product. The concept of *Value Engineering*, originally developed in the 1950s, emphasises the optimisation of a product's functionality to increase value by avoiding unnecessary costs and making functions more efficient. Value management approaches like this focus on realising the maximum value for the customer while minimising costs. The value of a product can also be influenced by various factors such as embodiment design, materials, and available manufacturing processes. Decisions regarding these factors can significantly shape the later product value even in the early phase of product

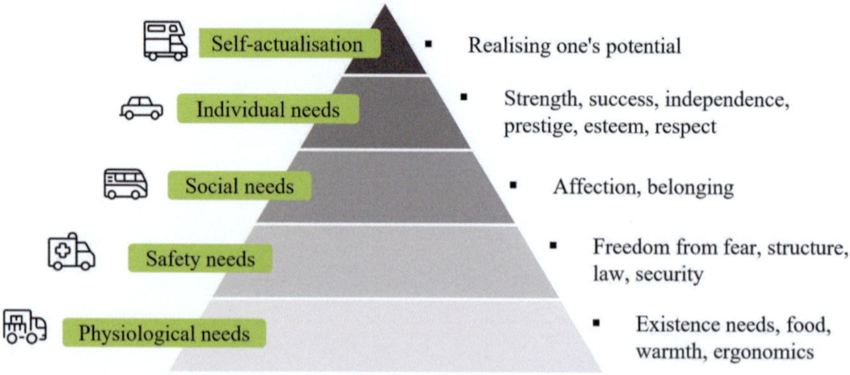

Fig. 1.4 Maslow's hierarchy of needs exemplified by different types of cars

development. For example, the selection of cost-effective but functionally equivalent materials or the optimisation of the production process can lead to a significant increase in value. Through interdisciplinary collaboration and continuous review of potential customer requirements, the value of a product and the achievement of critical sustainability goals can be actively influenced and optimised throughout the entire development process.

The hierarchy of needs often takes the form of a pyramid, as shown in Fig. 1.4, and is based on physiological needs, followed by safety needs, then social needs and individuality needs, extended by the need for self-actualisation.

Considering the product car against the background of Fig. 1.4, it can, if it is an ambulance or a lorry bringing food to consumers, help to satisfy physiological needs very sensibly. If it is a police vehicle, it can effectively support the fulfilment of safety needs. Cars meet social needs when they facilitate the commute to work or enable visits to family and friends. Individuality is realised in dealing with cars through brands, equipment, and other specific versions. Finally, there is the level of self-actualisation. Here, certain types of cars (motorhomes, convertibles, classic cars) can be classified, but also car races or the regular posing at popular excursion spots.

Depending on the usage and positioning of products in the hierarchy of needs, it is thus necessary and possible to argue in a differentiated way about the necessity and meaningfulness of these. In particular, the question must be asked how many resources, especially in affluent societies, are and must be consumed for individuality and self-actualisation, and whether there are no other solutions and ways to achieve increased life satisfaction in general. Less debatable, however, is the general benefit of technical products for meeting physiological basic needs, safety needs, and at least a canon of social needs. Here, only better solutions can replace what is established today. Since, according to Bertolt Brecht, food comes before morality, this leads to very different evaluations of technical achievements in a global population that has doubled to over eight billion people within 50 years.

1.3 Impacts of Products

In the context of the development of physical and non-physical products, the recording of economic and ecological impacts within a company is focused (see Chap. 5). Starting from a value or need debate, however, the question also arises about the social impacts of products. To answer this question, social sustainability must first be defined. Characterised as a "peaceful and fair coexistence" taking into account equal opportunities, education, and political participation, social sustainability can be defined. Specifically, these factors serve as a basis for recording the positive and negative impacts of a product on, in summary, the well-being of society in the form of, for example, Social Life Cycle Assessments. The complexity of recording these impacts and collecting the necessary data and information is evident in a content excursion into the "ethics of ecological responsibility". According to Hubka [6], man sees himself "as part of nature and the environment and is aware of his responsibility for the development and use of sustainable technologies and products". In reality, the actual assumption of social-societal responsibility by companies is not clear. While legal requirements, such as the Act on Liability for Defective Products (short: ProdHaftG), oblige companies as manufacturers, rules of social responsibility are less binding. Summarised as Corporate Social Responsibility (short: CSR), societal aspects, such as social benefits for employees, are also in focus alongside value-creating processes. However, the implementation and extent of these additional measures are not uniform and thus difficult to measure. In the following Fig. 1.5, CSR is placed in the context of a corporate sustainability view and illustrated.

According to the Scientific Advisory Board of the Federal Government, *Global Environmental Changes,* a resulting societal responsibility (see Fig. 1.5 "Corporate Social Responsibility") must be divided into three aspects and lived out as such for future generations. The three aspects include ecological responsibility as a culture of mindfulness, democratic responsibility as a culture of participation, and future responsibility as a culture of obligation [8]. This obligation is not to be

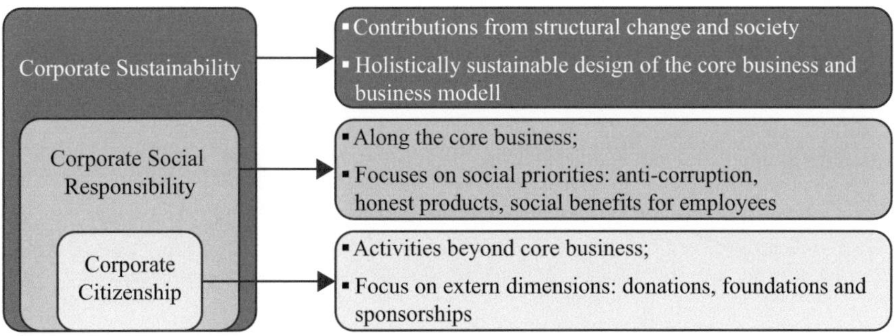

Fig. 1.5 CSR—Corporate Social Responsibility; own representation based on Schaltegger and Petersen [7]

understood as a restriction of the company, but rather as a task to also understand the consequences of actions from a social perspective. One way of dealing with these consequences is the principle of precaution. Distinguished into a risk and a resource precaution, measures can be stopped before implementation if they harm society or the environment. While risk precaution involves an assessment of potential dangers and damages of the respective products, for example, with the help of the method of failure mode and effects analysis (see Sect. 11.1), resource precaution aims at conserving available primary resources.

Methods for implementing social sustainability in product development often align with the needs and values of the users (see Sect. 1.2). The method IKIGAI, originating from Japan (eng. "iki" = life/"gai" = value), extends this approach to the meaning of life. The IKIGAI method can also be seen as an approach to realising a human-oriented product planning. Intending to develop products for both individual and practical need satisfaction as well as considering emotional, cultural, and social desires, IKIGAI combines the aspects of mission, profession, vocation, and passion as perspectives of product development [9] (see Fig. 1.6).

Below are some examples of socially-oriented approaches to product development:

- **Shared Economy**: As an economy of sharing, this approach aims at the communal use of goods through sharing, swapping, borrowing, or renting. With the aim of better utilisation of existing capacities, reducing absolute resource consumption, and increasing social contact, the approach of the so-called shared economy is visible in many everyday products. Whether as rental bikes, open-source software, like GIMP, or as a digital platform, like classified ads or Airbnb, sharing and exchanging physical and non-physical products is already common today.

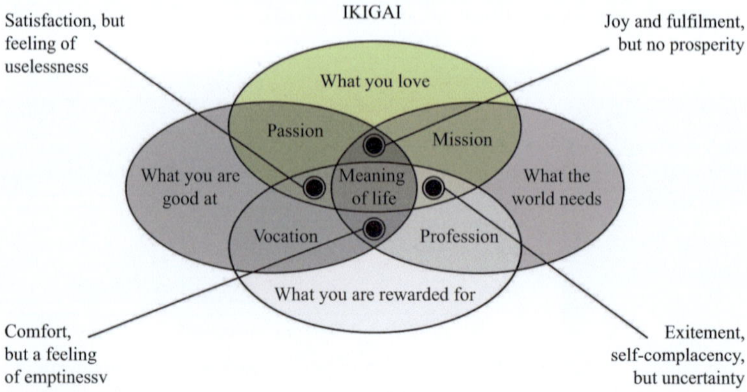

Fig. 1.6 Value-oriented product development through the IKIGAI method; own representation after Sato [9]

- **(Individualised) Medical Products**: From the perspective of patients and medical professionals, the development of individualised implants, for example, for hip bones or knee joints, is challenging both in terms of cost and manufacturing. Through the use of additive manufacturing technology and the development of algorithms and digital tools for individualising implants, the lifespan of implants can be increased, for example, by avoiding stress shielding.

Frugality: From an economic perspective, the development of frugal (i.e., simpler or modest) products is subject to the "creative omission" of functions or components. In other words, products that make do with the bare essentials and fulfil the sufficiency concept. In addition to the desire for reduction and simplicity of products, adapting innovative products to markets with less capital is necessary to satisfy these needs.

1.4 Challenges in Product Development

This book is about the development of sustainable or more sustainable products. From the perspective of development methodology, the question arises: What measures are required to minimise the negative impacts of production processes and the operation of products as much as possible and optimise their ecological and social impact?

Over the decades, many views on the human activity of product development have been established. These range from rather descriptive views, which describe development as a highly iterative trial-and-error process or highlight alternating analysis and synthesis phases, through very team-oriented perspectives, views that discuss various roles in the development process, to the process flow-oriented views that are very popular in management. Figure 1.7 shows some examples in this context.

The six views on the product design process shown in Fig. 1.7 are not complete, but they cover a broad field of development methodological approaches. While (a) divergence and convergence as well as (b) waterfall follow a stringent direction to achieve the goal, the (c) cyclical approach and the (e) complex-linear procedure allow iterations and planned revisions in product development. In addition to these structured approaches, it can be particularly useful in exploratory development projects to be guided by the "chaos". In the form of (d) discovery, the next steps in development arise from the development itself. This also corresponds to the SCRUM approach. The matrix approach is particularly suitable for large and complex development tasks that require parallel collaboration of multiple teams.

New approaches, such as working with innovation cells, start-ups, think tanks, or open innovation platforms, i.e., agile methods, are necessary to achieve real progress according to the "Out of the Box" approach. These advances often occur in close connection with scientific education and the latest research results.

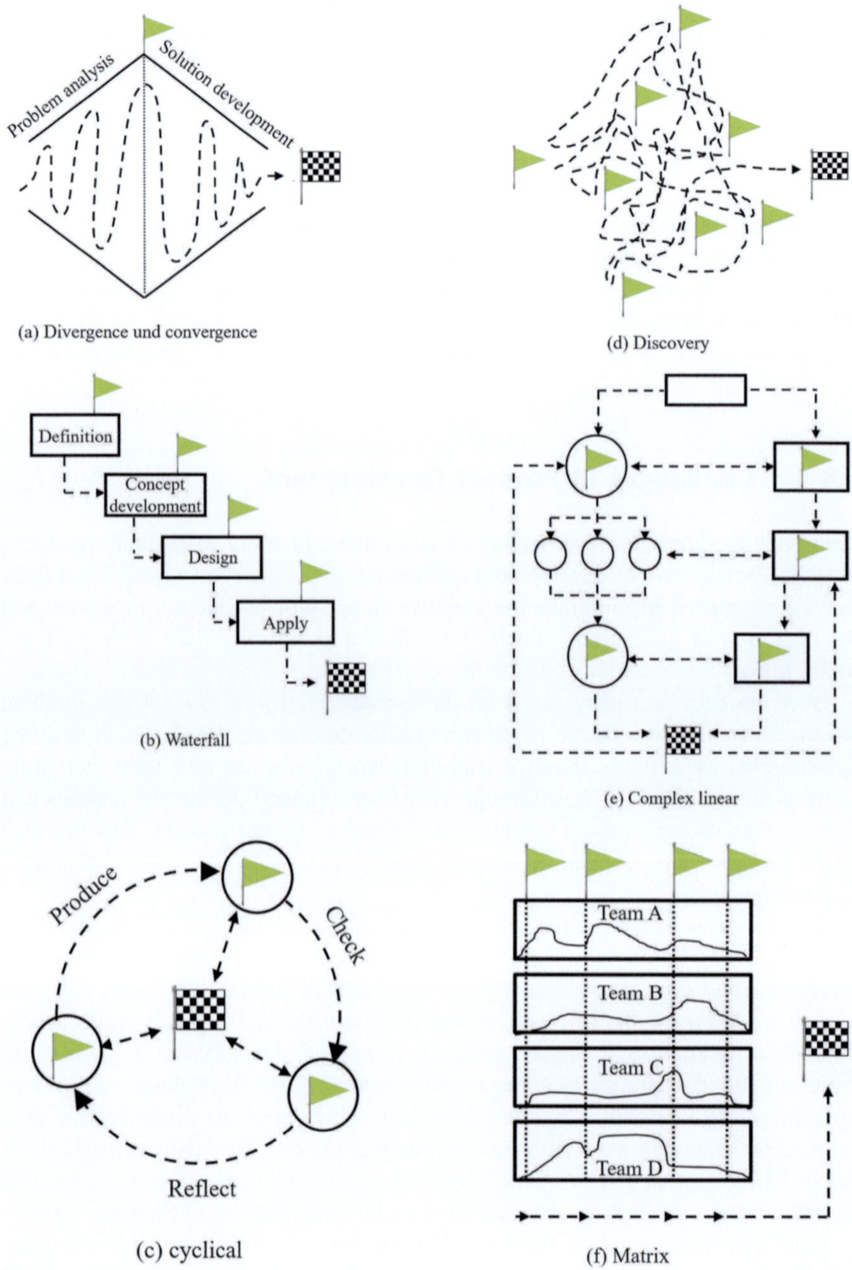

Fig. 1.7 Views on the development process

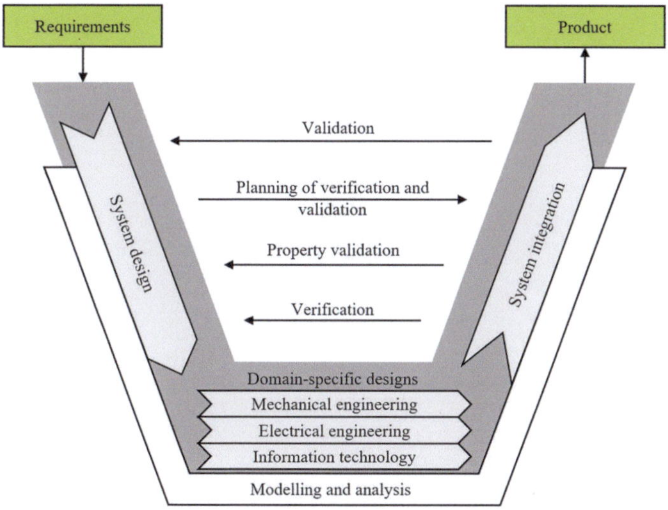

Fig. 1.8 V-diagram of the development of mechatronic systems according to VDI 2206 [10]

Explicitly in the German-speaking area, but also internationally, there are numerous works on the procedural approach to development and associated development processes. Commonalities between the different methodological schools are formulated in various VDI guidelines. Figure 1.8 shows the process for developing mechatronic systems according to VDI 2206. The starting point of the process is the specification of the system at the top left end of the model. Part of this specification is the business model, which summarises the company's fundamental goals, strategies, and measures. From this point, domain-specific challenges and subsystems are solved in a step-by-step approach, which are then gradually verified and assembled against the specification in the right section of the model. Finally, the overall system is verified and validated in its real application, among other things, to secure all properties.

The V-Model in its original form describes the development from the "big" to the "small" and back to the "big". Specifically, this means that certain requirements are transferred into an overall system design and then designed in detail in the individual subdomains. In a final system integration, the product is finally elaborated.

Without wanting to go into the many other facets of development processes and their implementation in different industries—see, for example, VDI 2221 (Task clarification, Concept development, Design, Detail design) [11]—a statement should be formulated at this point. It should be noted that these processes are also fundamentally suitable for structuring the development of products with special consideration of sustainability aspects. Special features arise concerning the new methods to be developed and used, which will be examined in detail in

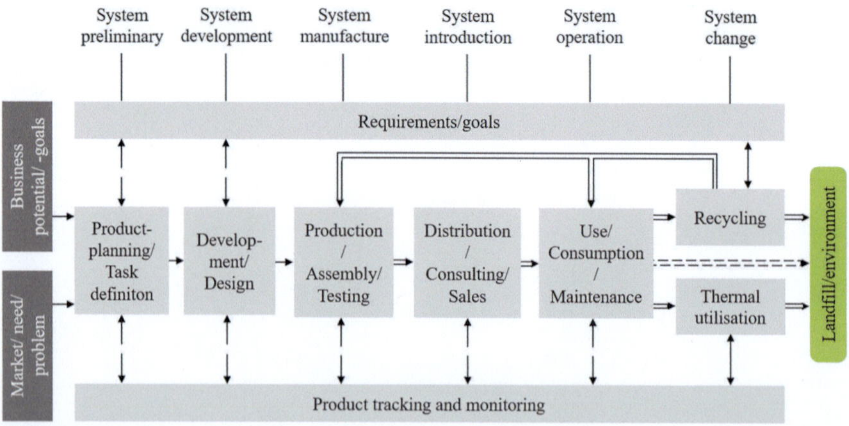

Fig. 1.9 Product life cycle according to VDI 2221 [11]

the further course of this book. Based on this, the subsequent use of these methods in various process phases will also play a crucial role.

Another important aspect from the point of view of methodology for considering the sustainability of products is the consideration of the entire product life cycle. In this context, the development, production, distribution, usage, and end of product life are particularly noteworthy. However, this aspect is not new either. Figure 1.9 shows a picture from VDI guideline 2221.

The VDI 2206 (V-Model) and the VDI 2221 (Flow model) are two central approaches in product development that complement each other in their structure and application. The V-Model from VDI 2206 describes a systematic, iterative development process. It emphasises the importance of a clear goal definition and a structured verification and validation at every point in the development. The VDI 2221, also known as the flow model, focuses on the methodical embodiment design of components and mechanical constructions by providing an iterative approach to solving complicated tasks. Both approaches offer the starting point of a targeted product development, with the V-Model primarily ensuring the quality and traceability of the development process, while the flow model allows for flexible handling and continuous adaptation of development activities. In combination, they support a holistic and efficient product development that enables both structured processes and creative solutions.

For the following understanding, this process representation of product development is also followed, as it allows the assignment of the methods described below to specific development phases. However, it should be noted that existing development organisations and processes are often only limitedly suitable for innovative solutions and projects, as required in the development of sustainable products. The efficiency of these existing processes lies primarily in the development of cost-effective and high-quality product generations and less in the elaboration of ecologically, economically, and socially more sustainable aspects.

When considering sustainability in product development, there are various special features in the use of methods. This includes the selection and application of business models that provide maximum performance with minimal resource use. Methodically, the so-called product service systems and concepts of the "Shared Economy", which will be discussed in more detail in Chap. 2, offer interesting options. Also, the question of how intrinsic motivations influence consumer behaviour is to be considered in this context.

Challenges also arise for the specification not only from wishes and self-defined goals, but also from legal frameworks and guidelines that must be complied with and thus are brought into the development as binding requirements. Associated methods are examined in more detail in Chap. 3. This field of work is now extraordinarily extensive due to national and international legislative initiatives. It is also developing very dynamically and deserves special attention. After all, disregard of these demands leads to total failures of development projects or, if the Products are only in the market, leading to significant subsequent costs.

The early stages of product development, particularly the concept creation phase, are emphasized here. An associated illustration by McAloone [12], which shows the relationship between time (x-axis) and the environmental impacts (y-axis) of the development project, clarifies that about 80% of the total environmental impacts of a product are determined during the concept phase. The graphical representation in Fig. 1.10 highlights the engineering relevance of influencing the far-reaching environmental consequences of a product through sound technical decisions at an early stage.

By integrating sustainability criteria into the concept phase at an early stage, engineers can ensure that the environmental impacts are minimised throughout the entire product life cycle. This insight underscores the importance of thorough planning and a strategic approach in the initial development stages. Technical decisions made in this early phase have significant impacts on resource efficiency, emission reduction, and the overall environmental balance of a product. The visualisation in Fig. 1.10 serves as a striking reminder that technological innovations and environmentally conscious embodiment design must begin in the concept phase. This emphasises the responsibility and potential of engineering to integrate and promote sustainable solutions from the outset.

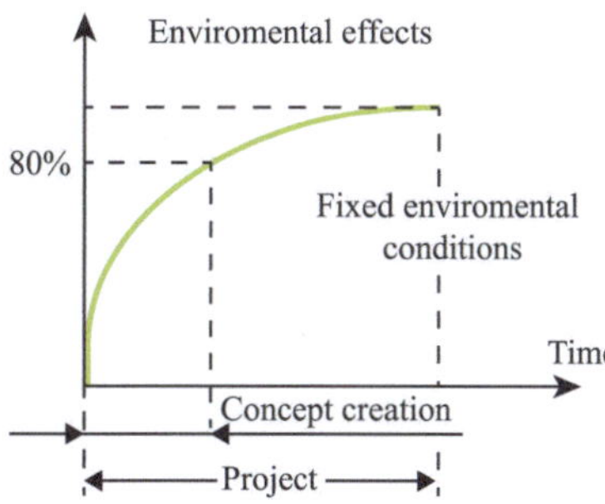

Fig. 1.10 Influence of product development on the environmental impacts of a product according to McAloone [12]

In development, it is also necessary to consider the various calculation approaches of the ecological footprint and their influence on the economic aspects of the product and its usage. This results in new evaluation standards for the promising selection of ideas and solutions. Crucial for the technical content of the development is the question of which natural scientific physical aspects can be effectively used as innovations. This raises the questions of which new paths must be taken and how the associated risks can be managed in the shortest possible development times.

Essential in the development of products is not only the "big hit", but also the concrete and detailed implementation. This is the only way to create systems that work throughout their entire life cycle. Therefore, concrete calculations, simulations, design guidelines, and principles that, for example, promote the recycling or repairability of products must be researched and made usable. These essential aspects and associated methods will be examined in more detail in the course of the book.

Ultimately, in development, it is important to ensure that products reliably function as long as they are supposed to. Here, product design planned with sustainability aspects in mind plays a major role.

The definition of the optimal product lifespan is also challenging. Too short lifespans lead to high resource consumption, while too long lifespans can result in disproportionate price increases and too high technological or technical innovation cycles.

Finally, we will try to balance what has been achieved and what is still open from a methodological point of view, based on the described process.

2 Systems and Business Models

Thinking in systems means considering the world. Chapter 12 contains numbers, data, and facts as well as sources that enable rational reflection on the necessity of further development and the derivation of business models from it as a global player.

For those developing a product, it is important to clearly define the framework of the respective development, or in other words, the system to be considered, to effectively carry out their engineering activities. A system is described by its system boundary, its inputs, and its outputs, according to Fig. 2.1, each divided into the three general sizes of technology: energy, material, and information.

Furthermore, a system can be characterised by the functional relationship between its inputs and its outputs. Depending on whether a system can only be described by the difference between inputs and outputs, i.e., by its effect, or also by its architecture, i.e., functional and structural design, we speak of a "black-box" or a "white-box" model of the system. The goal of development must always be to move from a "black-box" describing the desired effect to a "white-box".

Even in the technical environment, systems can be extraordinarily complicated and, for example, be networked with each other via the internet and interact. In this case, we then speak of systems-of-systems or cyber-physical systems. To make these manageable for development, it is important to structure them, that is, to specifically divide them into subsystems with their functions and subfunctions, and to standardise interfaces. These can then be controlled in development according to the approach shown in Fig. 1.8 (V-model).

Importance of System-Based Approaches

To make the previously described complex or complicated, and large systems feasible and manageable, systems engineering is suitable. An explicitly interdisciplinary

R. Lachmayer et al., *Methodology for the Development of Sustainable Products*, https://doi.org/10.1007/978-3-662-72442-2_2

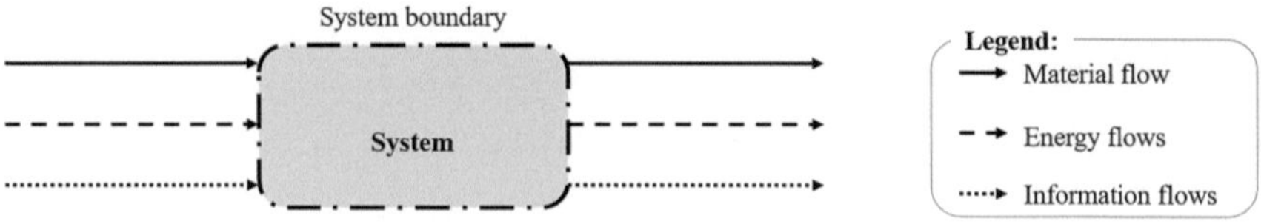

Fig. 2.1 System with system boundary and inputs and outputs

approach is pursued, which offers both processes and methods for the successful development of systems. This approach focuses on the early identification of customer requirements and necessary functionalities, while demanding a complete documentation of all requirements before the validation and embodiment design of a comprehensive solution begins. The solution formulated subsequently encompasses aspects such as system operation, cost and time planning, performance, training, maintenance, testing, production, and disposal. The concept of systems engineering takes into account both societal and technical requirements with the distinct aim of creating high-quality products that meet the needs of the users.

When discussing sustainability, the following systems are often mentioned from a societal perspective. These affect all areas of societal life and are closely interrelated with the use and development of technical systems and subsystems:

- Agriculture
- Food system
- Mobility system
- Energy supply system
- Health system
- Communication and information system
- Internal and external security system
- Education system
- Legal system
- Science system.

For example, the mobility system has impacts on the emission of pollutants into air, water, and soil (ecology), the costs for infrastructure and operation (economy), as well as accessibility and social integration (society). Similarly, the energy supply system and the education system shape both ecological aspects through resource use and economic and societal factors, such as equal opportunities and long-term economic stability. A sustainable technical system must therefore not only be efficient and resource-saving, but also take into account the social needs of society while ensuring economic profitability. The interdependencies between these areas require an integrated approach for each subsystem development to create long-term and holistically sustainable solutions.

The sections of this chapter will further elaborate on various aspects in the context of systems thinking and business models. Initially, the focus will be on the influence of business development by politically installed instruments such as laws, rules, or guidelines. The Renewable Energy Act (short: EEG) and certificate trading are cited and discussed as examples in this context.

The subsequent Sect. 2.2 will deal extensively with scenarios. After an introductory definition of scenarios in general, the scenario technique as a method will be presented using the example of a scenario for the development of sustainability awareness in a selected region of Europe. Subsequently, three different future visions will be developed, which should be understood as a framework for decisions on the portfolio.

In Sect. 2.3, various considerations on corporate strategies, portfolio management, and possible strategy pitfalls are discussed.

Finally, the focus in Sect. 2.4 will be on novel business models that are often discussed in the context of multidimensional sustainability. In this context, the method of the Business Model Canvas with an extension of factors of ecology, economy, and social aspects is presented as a possibility for the sustainability-oriented development of business models.

2.1 Legal Influence on Markets

Strengths and weaknesses, as well as opportunities and risks for system and business model development considering sustainability aspects, result from the shaping and further development of political instruments. There is the possibility of incentivisation, i.e., rewarding initiative and progressive behaviour, and the possibility of penalisation, i.e., in the form of levies and fee increases, and ultimately commands or prohibitions.

Some examples illustrate how strong the influence of the legislator on product development is, and at the same time, how strong the effects on a strategic product definition are. Assessing this influence correctly or even manipulating it through lobbying is essential in the early phases of product creation.

Through the Renewable Energy Act (short: EEG), feed-in tariffs for electricity from renewable sources have been guaranteed since 1 April 2000, which are to be borne by the consumer through levies. Each of the numerous amendments to the law, most recently on 22 May 2023 [13], significantly influences the profitability of investments and thus the order development of the renewable energy industry with its approximately 300,000 employees in Germany. Current extensions of the law focus on efficiency in the collection and processing of data and information. For example, the current conceptual design of the law (2025) makes the introduction of "smart" electricity meters for photovoltaic systems mandatory in order to provide a uniform data basis for the assessment of receiving feed-in tariffs and to control these depending on demand.

Emissions trading is an environmental policy instrument aimed at cost-effectively reducing the emissions of greenhouse gases such as CO_2 and its equivalents. Under the EU Emissions Trading Scheme (EU ETS), introduced in

2005, companies are required to pay for their CO_2 emissions. The system is continuously adjusted, for example, through price increases of emission certificates, most recently at the turn of the year 2024/25. In addition to this market mechanism, the EU relies on further measures, such as the *Fit for 55 initiative.* As part of this, a reduction of emissions by 55% by 2030 is being pursued. Furthermore, the EU plans to extend emissions trading to other sectors such as transport and buildings. In Germany, a CO_2 pricing system for sectors outside the EU system was also introduced. This additional pricing mechanism aims to further advance emission reduction and achieve national climate targets. The Federal Republic of Germany also supports the introduction of CO_2 compensation projects and promotes a national adaptation of emissions trading through investments in green technologies and renewable energies. The introduction and continuous adjustment of emissions trading are central elements of European and German climate policy, with the aim of fulfilling the EU's planned climate targets as well as existing and binding global climate protection obligations [14].

2.2 Scenarios

Scenarios, as hypothetical future images of a sector and its associated development path, provide orientation regarding prospective developments. The aim is to identify relationships between these factors over a future period based on systematically describable factors. The statements about these relationships can be both qualitative and quantitative according to Teich et al. [15].

In addition to the political developments described above, numerous other aspects and factors, which are subject to complex interactions, influence the future and thus the framework conditions for the products of tomorrow. Some of these factors are categorised and listed below.

- Economy:
  - Market
  - Legal framework conditions
  - Political influences and subsidies
  - Company-specific framework conditions
  - Fixed and variable costs
  - Availability
- Technology:
  - Standards and norms
  - Automation and processes
  - Innovations
  - Materials
  - Reliability and life cycles
- Ecology:
  - Biodiversity
  - Consumption

  - Emissions
  - Ecosystem feedback
- Social:
  - Demographic change
  - Changing world of work
  - Value change and individualisation
  - Urbanisation and changes in living space.

For the strategic orientation of developments and the design of future successful product portfolios, it is essential to gain as precise a picture as possible of the future market in the specific company context. A proven method for dealing with future developments is the so-called scenario technique.

Here, the frame of reference is first defined, and then five to ten relevant influencing factors are determined. These can be assigned different characteristics for the future, typically for a period of ten years.

From plausible combinations of the different characteristics of the influencing factors, various types of scenarios, usually at least three, are derived for the future (see Fig. 2.2):

- **Extreme Scenario (Optimistic)**: Most favourable future development
- **Trend Scenario (Continuation of today's situation)**: Nearly constant progression
- **Extreme Scenario (Pessimistic):** Least favourable future development.

The development of such future images allows the scenario technique to facilitate a discussion on how a company must strategically position itself to robustly respond to future developments. Based on the formed scenarios, companies can consequently be enabled to develop adapted strategies and work out goal-oriented measures.

According to the approach by Gausemeier for the development of scenarios, three steps in two different phases—the scenario field analysis and the projection development—are necessary. As part of the scenario field analysis, influencing factors originating from the global environment (including politics, economy, society,

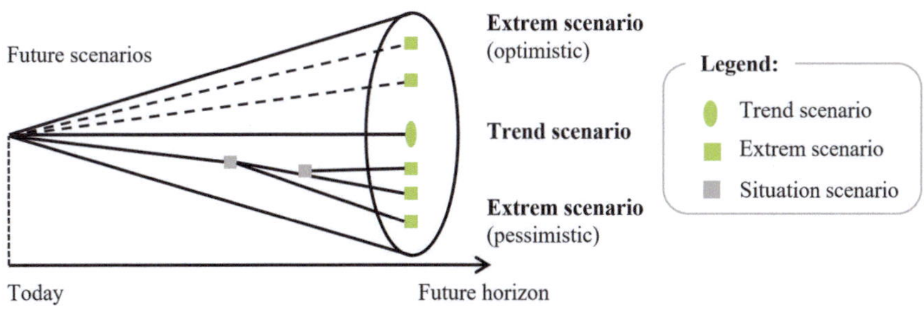

Fig. 2.2 Scenario technique; own representation based on Gausemeier [16]

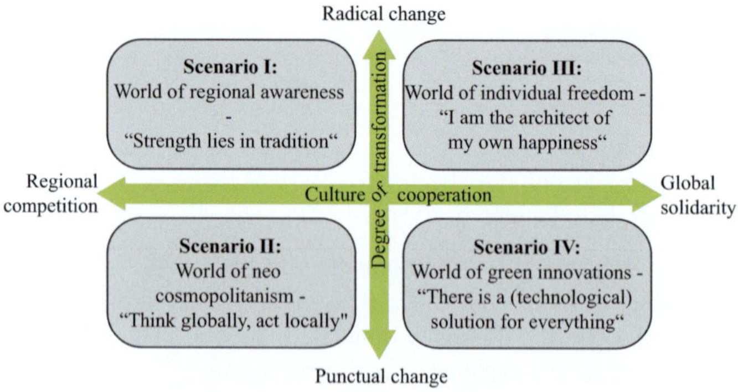

Fig. 2.3 Scenarios for the development of the South Tyrol region, according to Habicher et al. [18]

technology, or environment) on the considered company are identified. These influencing factors serve to describe the system. Based on these influencing factors, key factors are systematically determined, which describe the subject of investigation in a networked form. In the subsequent projection development, different development possibilities are worked out based on the key factors, which enable a characterization of the scenarios [17].

As a concrete example of the application of the scenario analysis, the result of a study by Eurac [18] is presented in the following Fig. 2.3. This study presents four possible scenarios for the development of the South Tyrol region based on investigations of the aspects of world, society, health, economy, environment, politics, and technology. Positioned in the field of tension of the culture of cooperation and the degree of transformation, a total of four scenarios are identified.

The Eurac study from 2020 presents a total of four scenarios for a possible development of South Tyrol up to the year 2030 and beyond. It is then checked whether and to what extent the defined sustainability goals of South Tyrol are achieved in the individual scenarios. It is based on a strategic foresight that considers various global trends and their impacts on South Tyrol. The analysis focuses on four equivalent scenarios, which show different development directions: a regionally conscious approach, a neo-cosmopolitanism, an emphasis on individual freedom, and a "green growth". Through expert interviews and the study of specialist literature, central themes such as climate change, urbanisation, and technological developments were identified. The scenarios visualise the possible societal, ecological, and economic changes.

The scenarios of the Eurac scenario analysis, according to Habicher et al. [18], differ in their central approaches and goals:

- **Regionally Conscious Approach (Scenario I)**: This scenario focuses on a strong regional identity and autonomy. It emphasises the protection of local resources, sustainable development, and the preservation of traditions, with a stronger regionalisation and the local economy and culture in the foreground.

- **Neo-Cosmopolitanism (Scenario II)**: Here, a cosmopolitan attitude is at the centre, intending to integrate South Tyrol more into global networks. The scenario promotes international cooperation, openness to diverse cultures, and a modern, globalised economic structure.
- **Individual Freedom (Scenario III)**: In this scenario, the focus is on personal freedom and self-responsibility. It promotes innovation, technological solutions, and the strengthening of individual life designs. The state plays a lesser role in this scenario while the self-realisation of citizens is in the foreground.
- **Green Growth (Scenario IV)**: Green growth relies on sustainable development that combines ecological and economic goals. It focuses on the expansion of renewable energies, environmentally friendly technologies, and an economic approach that takes into account both ecological and economic requirements.

Each scenario describes different ways in which South Tyrol could respond to societal and economic challenges, based on different priorities and values, and opens up new action options for product development, which should be robustly designed to deal with the impacts of these scenarios and their likelihood of occurrence. Many aspects of the scenarios can be transferred and adapted to other regions of Europe.

2.3 Corporate Strategy and Portfolio Management

The corporate strategy—combined with the vision and mission of an organisation—plays a crucial role in the alignment of the entire organisation. It defines the long-term goals and priorities of a company and thus influences the structure, processes, and allocation of resources in all areas. A key aspect in this context is portfolio management, which enables the control of product development. Portfolio management refers to the systematic selection, prioritisation, and control of projects and products that contribute to the realisation of corporate objectives. The use of portfolios supports the targeted implementation of investments and the optimal use of resources. This is achieved by ensuring an overview of all corporate activities and aligning them with the corporate strategy. The benefit of portfolio management is that it provides a structured basis for decision-making that takes into account both strategic and operational objectives. This allows companies to ensure that their resources are focused on the most promising and sustainable projects. Methods such as technology portfolios or portfolio analyses for eco-efficiency offer a structured approach to identifying and promoting sustainable technologies. These approaches enable the product portfolio to be aligned with the corporate strategy for ecological efficiency and market needs, thereby specifically promoting the development of more sustainable products. This not only links ecological and economic objectives, but also contributes to the long-term competitiveness of the company. Portfolios are used to locate various corporate activities in existing or resulting tension fields.

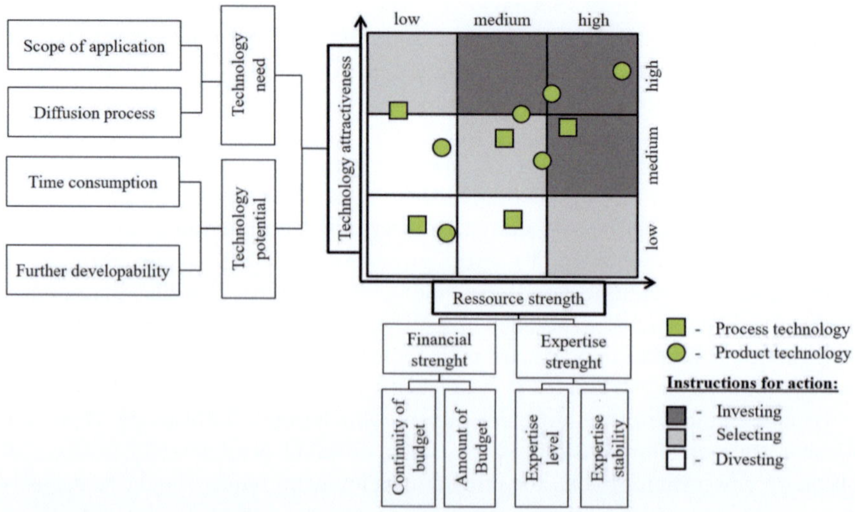

Fig. 2.4 Technology portfolio; own representation based on Pfeiffer [19]

The typical structure of a technology portfolio with the two axes of resource strength and technology attractiveness, as well as the action fields of investment, selection, and divestment, is shown in Fig. 2.4.

Technology portfolios are fundamentally based on the analysis of technology attractiveness and resource strength. Technology attractiveness describes the potential of a technology for future market success and innovation. Resource strength refers to a company's ability to mobilise resources (e.g., financial strength or expertise) to develop and support these technologies.

The possible actions within technology portfolio management can be divided into the following three types:

- **Invest**: Invest in technologies with high attractiveness and high resource strength to achieve competitive advantages.
- **Select**: Select technologies with medium attractiveness and/or resource strength to promote targeted further developments.
- **Divest**: Divest in technologies with low attractiveness and/or resource strength to free up resources for more promising projects.

This targeted control ensures that companies can focus their innovative power on the most promising and resource-saving technologies.

The technology portfolio is at the beginning of the consideration, as the long-term orientation of a company's technological base has a significant influence on the medium-term development portfolio. Technological changes can lead to fundamental upheavals in companies, as was clearly seen, for example, in the switch from incandescent bulbs to semiconductor technologies such as LEDs, or as the steel industry is currently facing, the switch from coal and gas to hydrogen. In the context of corporate development under sustainability aspects, both the demand-side and potential-side indicators are of particular interest for the assessment of technological attractiveness. For resource strength, the degree of mastery, the existing potentials, and the speed of action play essential roles. A distinction can be made between process technology, product technology, and, as described above, investing, selecting, and divesting.

With regard to product portfolios and the classification of development projects, the market-technology portfolio shown in Fig. 2.5 can be used as an example.

This portfolio aims to determine the attractiveness of a product under consideration from a market and innovation perspective (vertical axis) and the company's assertiveness based on competitive strength and technological performance (horizontal axis).

The size of expenditures or planned revenues can, as exemplified in Fig. 2.5, be characterised by the circle area when locating individual activities in the portfolio. This results in various areas where withdrawal from the market or no further investment is advised, areas where an indifferent situation should be reviewed, and highly attractive areas where leadership in both the market and technology should be sought. Interesting areas in the portfolio are also the lower left corner and the upper right corner. Here, the company's strengths lie either in technology or in the market. The other ability can be compensated, in addition to one's own efforts, also through joint ventures, licensing, or acquisitions.

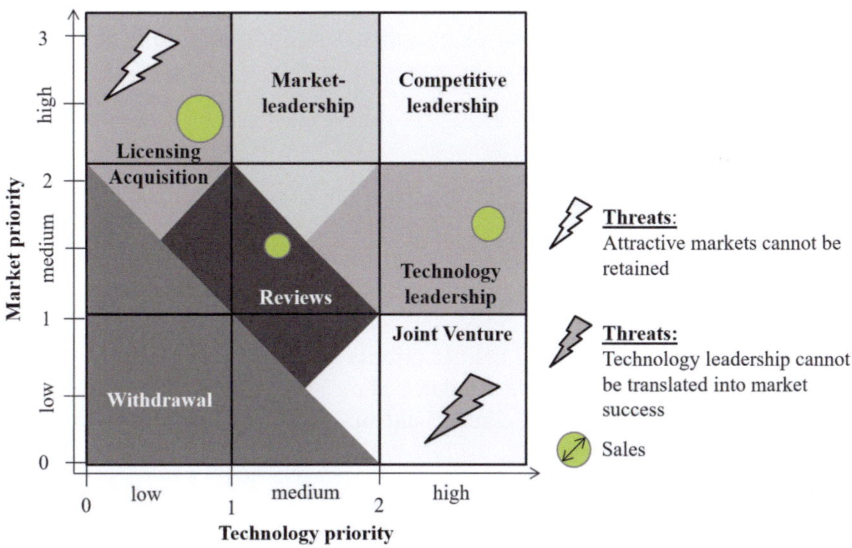

Fig. 2.5 Market-technology portfolio; own representation based on Albers [20]

The design of the market-technology portfolio is also always dependent on the life cycle of the market or product. Thus, the withdrawal area in newly emerging or growing markets is nonexistent, and other areas, such as niche strategy or technology presence, are more prominent.

The following will discuss the above through a practical example. The development of a technology portfolio for sustainable lighting, for instance, requires a classification of the various technologies in terms of their technology and market priority, as well as an analysis of potential hazards. The relevance of the lighting technology example is evident in the significant changes over the last 30 years in terms of technologies used. Starting from halogen and gas discharge lamps, as well as LEDs and lasers, subsidies, laws, and bans have favoured this shift. The function of technology portfolios is described in the following using this example.

Technology Priority:

- **LED Technology**: High technology priority is given as LEDs, due to their excellent energy efficiency, long lifespans, and decreasing production costs, which are the promising solution for the future. This technology is technically mature and has the greatest potential to replace traditional incandescent lamps.
- **OLED Technology**: Medium technology priority is given as OLEDs are promising, but currently face higher production costs and technical challenges in scaling. However, due to their flexibility and lower material consumption, they offer interesting potential for future applications.
- **Halogen Lamps**: Low technology priority is given as they are less efficient compared to LEDs and OLEDs. However, halogen lamps, like gas discharge lamps, played a role as a kind of transitional solution until fully sustainable alternatives were available.

Market Priority:

- **LED Technology**: High market priority is given as LEDs are now widely used, and the demand for energy-efficient lighting solutions and technological features, such as adjustable colour temperature, is growing. Consumers are also increasingly interested in sustainable and cost-effective alternatives, which promotes market acceptance and consequently market priority.
- **OLED Technology**: Medium market priority is given as OLEDs are gaining popularity, but currently have a higher price and are primarily establishing themselves in the premium segment and in specialised applications such as flexible displays and lighting.
- **Halogen Lamps**: Low market priority is given as they are being increasingly displaced by more environmentally friendly and efficient technologies due to their comparatively low energy efficiency and a heightened focus on sustainability aspects. They are already banned in many regions, such as Europe.

Hazards:

- **Technological Risks**: The development of LEDs and OLEDs requires continuous innovations in the fields of materials science and manufacturing technology. Technological setbacks or unsustainable production methods could potentially delay progress. In particular, production costs, quality risks and scalability could pose a challenge for OLEDs.
- **Market Risks**: Despite the high market acceptance of LEDs, the market could have been affected by regulatory changes or sudden technological breakthroughs of other technologies (e.g., with higher efficiency). The transition to more sustainable products could be adopted more slowly by consumers if initial costs are perceived as too high or the offered quality is too low.
- **Environmental and Resource Risks**: A potential hazard in the production of LEDs and OLEDs is the use of rare raw materials, which, due to their limited availability, pose a risk for resource scarcity. In addition, improper disposal or recycling processes, as well as the required electronics, can lead to environmental pollution. Efficient recycling processes would not only reduce the environmental impact but also the need for new raw materials.

A strategic portfolio that prioritises correctly and effectively minimises risks is essential for technological developments in a company. It is crucial for the long-term success of a product and consequently for the overall success of a business.

2.4 New Business Models

Business model development is primarily about essential core processes in a company, the respective corporate strategy, the expenditures for development, and the processing of orders up to delivery. Figure 2.6 shows the relationship between product development ("time to market") and realisation ("order to delivery").

The course of action for product development begins with a product idea, and consists of product and project planning, design, manufacturing planning, and production start-up. The product-specific design process itself is divided into the areas of task formulation, conception, conceptual design, and elaboration. Subsequently, it transitions into the area of product implementation.

The course of action for implementation or commercialisation of a product, as can be seen in the right half of the figure, essentially consists of six areas: procurement, production, logistics, operation, associated services, and recycling. Depending on the batch size, this process can be repeated many times for the same product.

The following Fig. 2.7 provides an overview of the processes described above in the context of corporate strategy. Starting from a company's mission statement, an environmental analysis consisting of competitive and market analysis, as well as the consideration of the company and its programmes, objectives can be formulated. These objectives are traditionally of a monetary nature (investments,

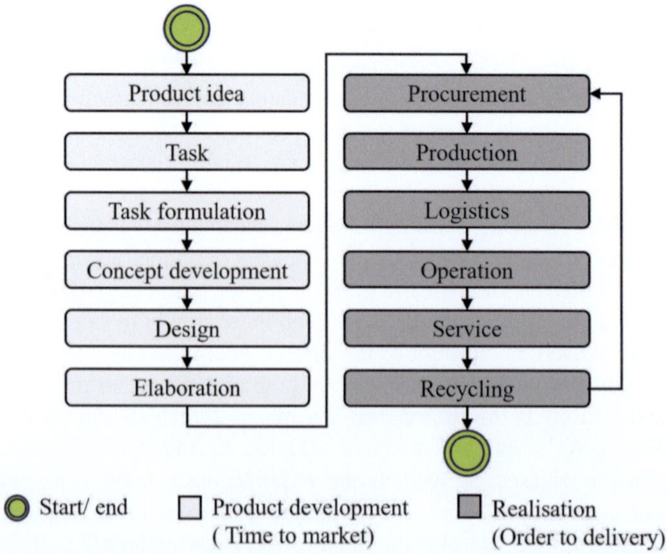

Fig. 2.6 Core processes in a company

revenues, returns, profits), but should actually, in the sense of a long-term corporate strategy, refer to so-called "Balanced Score Cards", i.e., balanced key figures. These key figures could then cover all aspects of sustainable development—social, ecological, and economic factors—such as competence and personnel development, the degree of usage of renewable resources, customer and market relationships, and supply chains.

In the context of developing corporate strategies, environmental analysis is a crucial component to identify existing external influences and opportunities, as well as existing risks that impact the company. This analysis includes not only the market and competitive situation, but also the interactions between the technosphere (the technical and economic systems created by humans) and the surrounding ecosphere (the natural, ecological systems). The interactions between these two spheres are of crucial importance, as technological innovations and industrial processes increasingly interact with ecological conditions. An example of this is the emergence of environmental impacts from industrial production, which affect natural resources and cause and accelerate climate change. Conversely, ecological changes can likewise influence technological development, for example, through the introduction of environmental regulations or the need to develop more sustainable production methods.

Companies must therefore not only monitor direct economic and technological trends, but also consider the ecological conditions that may optionally affect long-term competitiveness. A successful corporate strategy necessarily takes into account the interactions between the technosphere and ecosphere, promoting environmentally friendly innovations while simultaneously minimising the negative

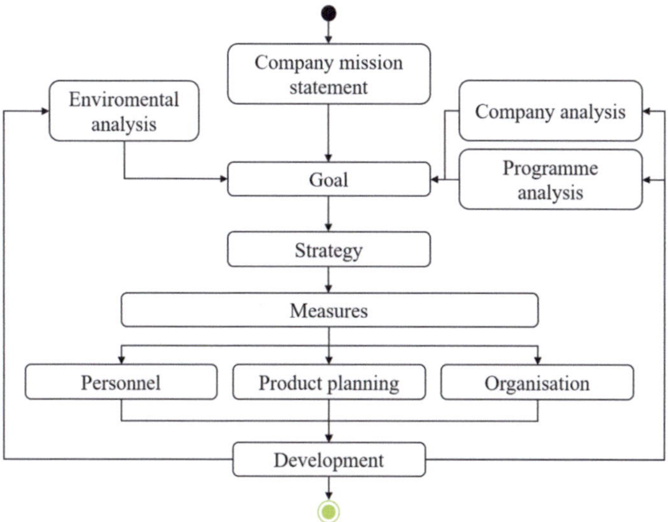

Fig. 2.7 Strategy development within a company

ecological impacts of its own activities. This requires a holistic and integrated view of technological developments and ecological goals, intending to develop, produce, and bring to market both sustainable and future-proof products and processes.

These defined objectives serve as the basis for the subsequent determination of strategy. From this, appropriate measures for personnel, organisation, and product planning must be derived and implemented.

In the context of developing sustainable products, innovative business models are sought after and purposeful. This added value often results from suitability for multiple uses and longer lifespans, which reduce the raw material and energy input for production. However, a change in customer behaviour is often also a prerequisite.

In the discussion of alternative business models, the following approaches are particularly relevant:

- Rental models (Renting)
- Hire purchase models with return guarantee (Leasing)
- Collective usage models (Sharing)
- Product-Service Systems (short: PSS).

The "Business Model Canvas" method will be introduced as a concrete example for the development of business model ideas. In the context of developing sustainable products, this concept, in addition to its original form, will be expanded to include aspects of sustainability. A graphical overview of the classic Business Model Canvas is shown in Fig. 2.8.

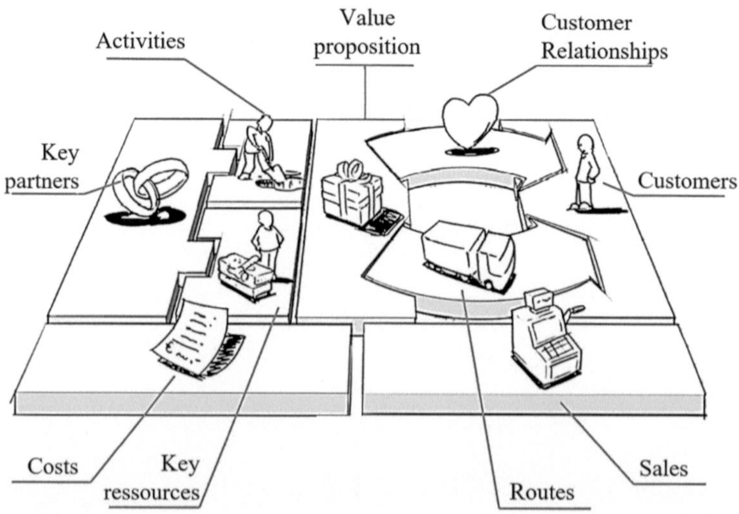

Fig. 2.8 Business model canvas: dimensions of business model development according to Osterwalder and Pigneur [21]

The model divides the necessary considerations into four dimensions: offer model, customer model, value creation model, and financial model. The elements listed in the figure, such as value proposition, customer relationships, the customer itself, revenue, channels, key resources, costs, key partners, and activities, are assigned to these dimensions. All dimensions of the business model influence sustainability. However, key factors for the assessment arise from value creation. As a rule, the relevant resources and energy-consuming activities that can be influenced are allocated here.

The integration of three-dimensional sustainability (ecological, social, and economic) into the entire Business Model Canvas requires an expansion of the classic model to encompass all relevant aspects of sustainable development. Below are three ways in which the Business Model Canvas can be comprehensively aligned with sustainability:

- **Sustainability as a central business model**: One option is to design the entire Business Model Canvas so that all components are directly aligned with sustainability. The company could commit to integrating sustainable practices in all areas (value proposition, customer base, activities, partnerships, etc.). For example, it could seek resource-saving or regenerative processes in the value chain and ensure that all suppliers and partners meet sustainable standards. In the cost structure, the company could transparently show how investments in sustainable technologies or fair working conditions lead to a reduced ecological footprint or better social impact in the long term. Particularly, functionally sustainable products, those that focus on sustainable business within the business model (e.g., heat pumps), can be mentioned.

- **Introduction of additional dimensions for sustainability**: Another way to integrate specific sustainability aspects would be to expand the Business Model Canvas with special components that explicitly address the three dimensions of sustainability. This could be done, for example, by adding an "ecological value" and/or a "social value" as separate components. These components would focus on the ecological and social impacts of corporate activities and ensure that these dimensions are separately integrated into the corporate strategy. In the customer segments area, an additional component for sustainability-oriented target groups could be introduced, which are particularly interested in environmentally friendly or socially responsible products (e.g., bicycle frames made of wood).
- **Triple-Bottom-Line approach**: Another expansion could be the introduction of the Triple-Bottom-Line approach (short: TBL) throughout the Business Model Canvas. This approach considers not only financial profit (economic dimension) as a success criterion but also the ecological and social impacts of the company. The Business Model Canvas would then be adjusted to consider three central goals: Profit, Planet, and People. Every decision in the Canvas would be checked to see if it makes a positive contribution to these three areas. For example, the partnership strategy could be adjusted so that the company only works with partners who meet both social and ecological criteria, thereby ensuring a sustainable impact along the entire value chain (e.g., vertical greening/planting for food cultivation in urban areas).

With these extensions, the Business Model Canvas becomes a comprehensive strategic tool that can enable companies to align their business models along the three dimensions of sustainability and subsequently create not only economic but also ecological and social added value.

A further core question to be answered could therefore be how to optimise the offer model and customer model in the context of business model development so that the greatest possible market performance can be achieved with as little material effort as possible.

In principle, this can be achieved by focusing more on fulfilling a promise of benefit rather than selling hardware. A comparatively stronger emphasis on the Product Service System concept usually leads to less material use and, with good maintenance, to a longer usable lifespan and higher reliability. From the perspective of the Business Model Canvas, the examination of product-service systems affects almost all elements. For example, the customer relationship changes from an interaction focused only on product sales to a comprehensive relationship spanning the entire product life cycle between the producing organisations and the users. The nature of this relationship depends on the individual concept of the product-service system.

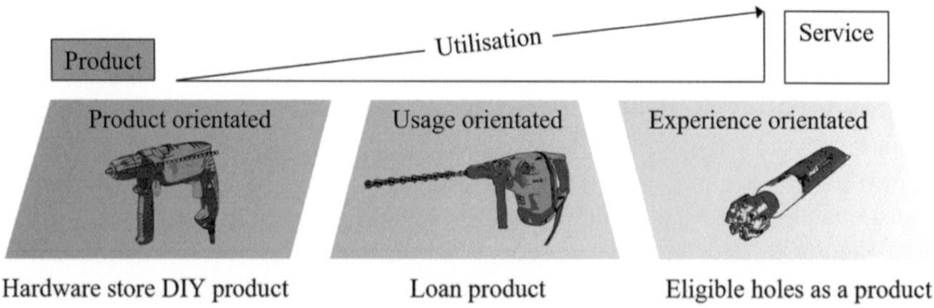

Fig. 2.9 Different manifestations of product-service systems depending on utilisation

The following Fig. 2.9 illustrates this approach or the concept of the product-service system using the example of variously designed drills.

While the small, inexpensive machine is usually bought and discarded when defective, there are providers in the middle segment who provide machines and services. In the right area of the graphic, where machine investments are very high, only services, in this example, the finished drill holes, are sold. Using a drill as an example, the PSS model can be implemented as formulated in the following paragraph to specifically address aspects of sustainability:

Instead of selling a drill, a company can offer a leasing or rental model in which the drill is regularly maintained, repaired, and equipped with new components as needed. This extends the product's lifespan and reduces the need for resources to manufacture new machines, leading to a reduction in the ecological footprint. In a shared economy model, drills could be rented to end customers for a limited period, meaning that the machines are only produced when needed. This minimises the need for surplus devices, as multiple users can share a single machine. This optimises both resource usage and energy consumption during the usage phase. The company can take back the old drill, recycle it, and use the raw materials for new devices. This reduces waste and promotes a circular economy, where valuable materials and semi-finished products are reused.

The utilisation of products plays a central role in the context of product-service systems, especially when it comes to maximising customer benefit and product efficiency. The example of cars illustrates that two factors are crucial for wear and tear: the mileage and the age of the vehicle. Mileage primarily affects mechanical components, as it is associated with tribological wear mechanisms such as abrasion in the engine and gearbox. The more a vehicle is driven, the more these stresses occur. In contrast, the age of the vehicle mainly affects material-related ageing processes—such as the embrittlement of plastics or the weakening of seals—which can negatively impact security and functionality in the long term. Product-service systems like car sharing offer a solution: shared usage significantly increases vehicle utilisation. This means that a vehicle experiences higher usage in a shorter time, improving the benefit per unit of time and per invested resource. At the same time, this more even usage avoids extreme strains on individual vehicles, which in turn reduces wear and tear. It also increases economic efficiency: vehicles spend less time idle, which improves the return on investment for providers and, at the same time, offers an attractive service for users. Overall, the car-sharing approach strikes a balance between intensive usage and controllable wear and tear, making the considered product-service system both ecologically and economically sustainable.

3 Compliance, Rules and Guidelines

Laws, rules, guidelines, and standards have influenced product development since people began trading goods with each other. On the one hand, these can act as strong drivers of innovation, enabling efficient industrial value creation in corporate networks. On the other hand, they regulate uncontrolled profit at the expense of others, society, or the environment. They also contribute to the security of all and the protection of nature by containing regulations and specifications that minimise the risk associated with technology and its implementation. This chapter will focus on this relationship, with particular emphasis on aspects of sustainability. The following sections will illuminate and elaborate on three essential aspects:

- Compliance, ensuring adherence to rules within the company;
- Laws and rules, which bind the responsibilities of companies;
- Guidelines and standards, which define product and process development.

3.1 Compliance and Reporting

Compliance refers to the reliable adherence to all rules within a company, whether they are internal guidelines or legal regulations. With the help of compliance management, companies can ensure that both management and employees behave in accordance with the rules. This affects the corporate strategy, the product and production programme, the organisation, the locations, the reporting, and even things like the "Code of Conduct", the company's own "Commitment", and the basic interaction with employees. The term "Compliance Management" is still relatively modern, although it describes a concept that should fundamentally be self-evident: the obligation of companies to comply with applicable legal regulations and rules. The fact that this is not quite as simple is due to the fact that laws are constantly

R. Lachmayer et al., *Methodology for the Development of Sustainable Products*,
https://doi.org/10.1007/978-3-662-72442-2_3

evolving and subject to political dynamics. Additional challenges arise from the fact that compliance, in addition to complying with legal regulations, also includes adherence to company-specific rules with content differences.

The organisation of compliance management can vary in complexity depending on the company. It can generally be divided into four aspects:

- Informing, with aspects of transparent guidelines, training, and advice;
- Content and operational development of reports and processes;
- Conducting audits, reporting, and identifying risks and problems;
- Implementation and resolution of problems, taking into account local legislation and culture.

The interplay between compliance and the sustainability goals addressed at the beginning of this book is characterised by interdisciplinary cooperation and an active linking of various fields of action, goals, and strategies. The necessary transparency towards stakeholders is an important aspect of compliance, with an increasing number of companies interested in meeting sustainability requirements. There is also an increasing legalisation of non-financial topics or the ecological and social dimensions of sustainability.

As part of compliance, various metrics are monitored that address both economic and socio-ecological aspects. Key socio-ecological indicators include compliance with environmental regulations (e.g., emission levels, waste management), working conditions (e.g., workplace accidents, fair work practices), and ethical procurement practices within supply chains. Other metrics concern compliance with anti-corruption policies, as well as the sustainability of products and the company's CO_2 equivalent accounting. These metrics contribute to taking on both ecological and social responsibility and ensuring that the company meets ethical and environmentally friendly standards. They also enable the company to monitor its compliance and implement sustainable practices throughout the entire value chain.

The attempt to measure the socio-ecological impact leads in the compliance context to the four-level model of corporate management, which is depicted in the following Fig. 3.1.

The four levels of corporate management can also be distinguished in the context of product development: operational, tactical, strategic, and normative. On the operational level, specific development projects are carried out, with technical details, schedules, and resource management at the forefront. This level is about achieving short-term goals and optimising production processes. The tactical level deals with the planning and coordination of product development strategies over medium-term periods, such as adapting product lines to market requirements. Decisions are made at this level that increase efficiency and ensure the marketability of the products. The strategic level is focused on long-term objectives, such as the development of innovative products, strengthening the competitive position, and market diversification. Here, the product portfolio is aligned with future trends, and the company is positioned for upcoming challenges. The normative

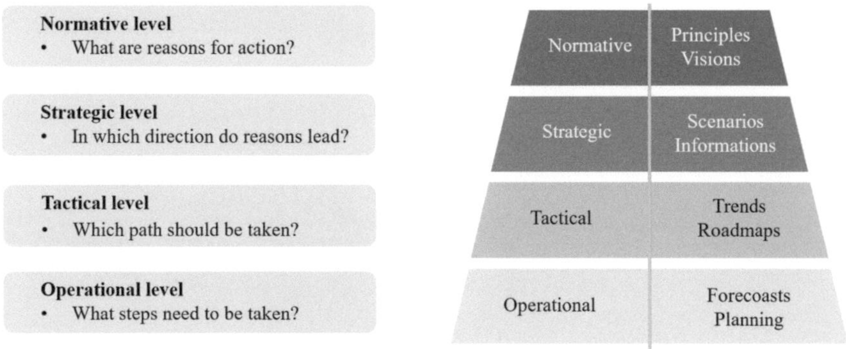

Fig. 3.1 Levels of corporate management; own representation based on Dyckhoff and Souren [22]

level includes corporate values, ethical guidelines, and fundamental principles that guide the company in product development, particularly concerning sustainability, quality, and social responsibility. These four levels work together to ensure a coherent and future-oriented product development.

While economic aspects are today aggregated across all corporate levels through business planning tools, this is usually less pronounced for the socio-ecological dimensions of sustainability—but would be necessary for successful reporting and corresponding target focus. Compliance management can provide structured support here, also in combination with the balanced scorecards discussed in Chap. 2.

The following will present selected examples relevant in the context of compliance and reporting, as well as monitoring and reporting, with a particular impact on the development of sustainable products. The selection does not claim to be exhaustive; the list merely provides a rough overview of how various regulations can influence the framework conditions for sustainable development.

DIN ISO 26000: Guidance on Social Responsibility of Organisations

Beyond the economic compliance topics, DIN ISO 26000 provides suggestions in the form of a holistic reference framework for dealing with social responsibility in the company. The idea behind it is to achieve greater acceptance for the organisation in the local environment through socially responsible action, to promote trust-building with contractual partners, and to have a positive effect on employees. Principles that are not always taken for granted are described as follows:

- Accountability
- Transparency
- Ethical behaviour
- Respect for stakeholder interests
- Respect for the rule of law
- Respect for international standards of behaviour

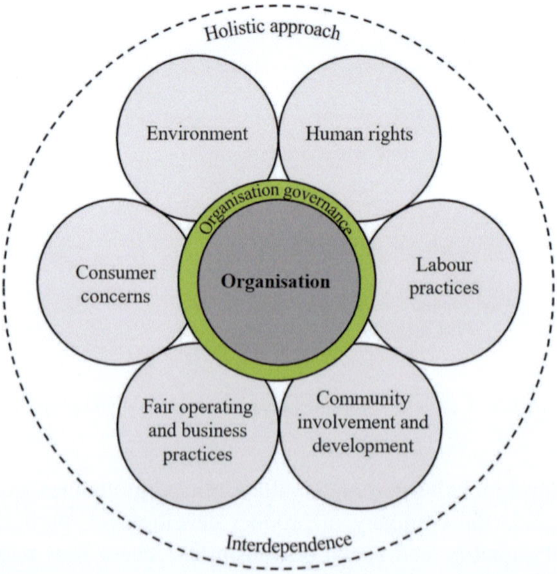

Fig. 3.2 Core topics of social responsibility according to DIN ISO 26000 [23]

- Respect for human rights.

The level of detail of DIN EN ISO 26000 is quite concrete and includes about 600 requirements in detail, some of which are presented in a condensed form below (Fig. 3.2).

EMAS: Eco-Management and Audit Scheme

EMAS is a fundamentally voluntary concept and scheme for implementing measures for systematic resource saving as recommended by the Federal Ministry for the Environment (short: BMU) [24] based on ideas from the European Community from 2014. EMAS aims to optimise company-specific environmental performance, develop and then establish systematic and internally developed environmental protection. The result should be compliance with all environmental regulations and public documentation of all environmentally relevant activities and data. The activities and measures to be carried out according to EMAS are listed in the following enumeration:

- Task clarification and preparation
- Define the guiding principle
- Develop an environmental programme
- Implement the system
- Conduct internal audits
- Create an environmental statement
- Have an external audit conducted
- Registration in the EMAS register.

According to the scheme, there are also economic opportunities for participating companies. The systematic approach can identify resource savings early on, while legal regulations are checked and legal certainty is ensured. Another advantage arises from the EMAS environmental statement. By presenting this statement, EMAS-compliant companies are able to credibly and verifiably advertise their environmental performance and impacts to the public. In summary, it can be stated that EMAS can make a significant contribution to sustainable business practices [25].

GRI: Global Reporting Initiative

The GRI is a non-profit foundation with a significant focus on a multi-stakeholder approach based on the ideas of the United Nations environmental programme. In contrast to the previously described EMAS, the GRI is primarily a globally recognised guideline for sustainability reporting. The main goal is to systematically, uniformly, and transparently support companies and organisations in the creation of sustainability reports by defining and issuing guidelines. The GRI covers all aspects of reporting on economic, social, and environmental indicators. The main tool for implementing this concept are the GRI standards and guidelines, which define a modular, structured, and user-friendly system for sustainability reporting. As a result, reports are comparable through standardised modules and may also be combined with the respective locally valid sustainability codes of the countries [26].

EPD: Environmental Product Declaration

An EPD, or Environmental Product Declaration in English, represents an environmental declaration of a selected product. It is created based on internationally valid standards for environmental labelling, such as ISO 14025. With the help of an EPD, products or services are described in terms of their environmental impacts, based on the entire life cycle, on the basis of life cycle assessments. With the help of the information obtained, targeted comparisons can then be made to decide between products or services of the same function. The basis for creating an EPD is objective, independently checked, and verified, as well as quantitative data from eco-balances or material balances. The EPD, similar to the well-known CE mark, is an environmental label, but in this case, it is voluntary. The neutral, factually based, and scientifically proven characterisation without the addition of an assessment is of central importance. An EPD is not to be understood as a certificate with certain requirements for a product, but rather as a uniform evaluation form for classifying the sustainability of different products. The EPD is mainly used in the selection, declaration, and assessment of building products [27].

SCI: Sustainability Compliance Index

In addition to legal compliance and the consideration of standards and guidelines, there are many other design spaces for improving the sustainability and especially the security of a product when developing products. In reference to the previously presented content on internal compliance measures, the SCI forms a further concretised addition. The area of monitoring and reporting plays a decisive role here. In the course of such evaluation procedures, various aspects of sustainability can be specifically addressed and detailed.

The SCI is a qualitative rating scale that serves decision-making within product development towards more sustainable products. The SCI's property catalogue compiles selected decision-making characteristics according to the phases of the product life cycle and the qualitatively descriptive values assigned to them. Table 3.1 shows excerpts from this; a detailed list can be found in [28].

The numerical values in the table each indicate the scale value of the SCI. An SCI value of nine describes the best possible level of strategic sustainability, a value of one the lowest level. An SCI value of zero indicates that too few or no information is available to create a value; in this case, additional investigations are necessary.

Table 3.1 Excerpts from the SCI property catalogue according to Hallstedt [28]

Production: Production by suppliers of sub-components and materials, as well as production of products in the company itself	Sustainability aspects considering sustainability principle 1		Sustainability aspects considering sustainability principle 2	
	9	(i) Only materials that do not emit metals are used, and all waste metals are recycled into pure fractions (ii) Only renewable energy sources are used in the production processes, and the energy usage has an efficiency of 2%	9	No emissions and waste products from production sites (including suppliers) contain substances from the SIN list
	1	(i) No recycled materials are used, and/or it is not known to what extent emissions and metal waste arise from production (ii) Only fossil fuels are used in production	1	Emissions and waste from production sites contain substances from the REACH candidate list
	0	(i) It is unknown whether and how much recycled material is used. No knowledge about the amount of emissions and scrap from production (ii) It is unknown whether and how many renewable energy sources are used in the production processes and whether the energy use in the production processes is efficient	0	It is unknown whether there are emissions and waste products from production sites that contain substances from the REACH candidate list

In addition to the inventory using the matrix shown above (see Table 3.1), the SCI can also look up guidelines for embodiment design or define strategic goals. A comparison of variants is also made possible by an appropriate assignment of properties.

In conclusion, it can be stated that even if the individual possibilities of sustainability reporting described here are each only described as recommendations or possibilities, companies must nevertheless create and publish environmental reports. These vary in quality depending on location, industry, history, etc. However, it can be assumed for the future that the respective rules will be further specified or tightened.

3.2 Corporate Responsibility Laws

Laws regulate and order in a legally binding manner. In laws on corporate responsibility, very general understandings are regulated, such as in the Act on Liability for Defective Products or the Electrical Equipment Regulation. Specifically, this means that processes, for example, in the Circular Economy Act and the Hazardous Substances Ordinance, are defined bindingly, or that in other laws, such as the laws to avoid harmful substances, there are very specific prohibitions.

As already mentioned in the previous section, selected and relevant examples in the context of legislation with a particular impact on the development of sustainable products will be listed below. The laws described, as previously mentioned, do not claim to be complete concerning all rules to be considered by a company.

ProdHaftG: Act on Liability for Defective Products

The Act on Liability for Defective Products regulates the obligation of a product manufacturer to be liable in the event of death, injury to the body or health, and property in the case of a product defect. The law applies in the Federal Republic; other countries may have similar laws. The manufacturer, in the sense of the law, is whoever has produced the final product, a raw material, or a part product. Anyone who identifies themselves as the manufacturer by affixing their name, brand, or other distinctive mark is also considered a manufacturer.

Errors that lead to liability are:

- **Manufacturing defects** (Deviations of the product from the standard of the product series, e.g., if food is contaminated with foreign bodies or defective components, wrong material qualities, or changed manufacturing parameters).
- **Design defects** (Designs that deviate from the state of the art and/or applicable standards and rules, e.g., if safety-relevant dimensions were executed incorrectly, undesirable materials were incorporated, or potential hazards were underestimated).
- **Instruction errors** (Faulty operating instructions regarding the manner of use. However, it is not considered an instruction error if general knowledge can be assumed, such as the fact that cats should not be dried in the microwave) [29].

CE Certification

The CE certification, loosely translated as "European Conformity", is a widely used and binding mark to ensure a certain product quality. Unlike the previously described ProdHaftG, this type of certification is not a classic law, but rather a legal requirement, which is anchored in many EU directives. This certification also has Europe-wide validity, unlike the ProdHaftG, and is therefore used as a standard throughout the EU.

By affixing or displaying the CE mark, the manufacturing company associated with a product assures that the respective product meets all EU-wide requirements for security, environmental protection, and health protection, and has been tested by appropriate bodies for these factors and has undergone a prescribed conformity assessment. With the CE mark, the company thus assumes full responsibility for the (EU) conformity of a product. In this context, it expresses that it knows specific requirements for the products it distributes and that these products meet these requirements.

For various products marketed in the EU, there is a marking obligation in the context of CE certification. The obligation to mark applies, for example, to medical devices, machines, electrotechnical products, toys, or telecommunications equipment [30].

ElektroG: Electrical and Electronic Equipment Act

The ElektroG implements the European WEEE Directive (*Waste of Electrical and Electronic Equipment*) 2012/19/EU into German law. This includes the regulation of the placing on the market, the return, and the environmentally friendly disposal of electrical and electronic equipment. The law first came into force in 2005 and was supplemented in 2015 (ElektroG2) and 2022 (ElektroG3).

The main aim of the ElektroG is to protect health and the environment, particularly through contact with harmful substances from electronic devices. The law also obliges the respective manufacturers to take back and collect old devices in a legally compliant manner. In this context, the ElektroG aims to conserve natural resources and reduce their use. Key components of the law are the efficient and high-quality recycling of electronic waste and an overall avoidance of waste. By complying with the ElektroG, manufacturing companies are obliged to take responsibility for the entire life cycle of the products they manufacture and put on the market. The law considers various areas of the aforementioned life cycle of electronic products. This includes, for example, guidelines for product ssdesign, labelling, return concepts, disposal strategies, certifications, recycling practices, information obligations, and many more [31].

KrWG: Law on Closed Cycle Management and Waste

The term "waste" includes all substances or objects that their owners want or need to dispose of. The Closed Cycle Management and Waste Law regulates the comprehensive circular economy and environmentally friendly disposal of waste. The following regulations are included:

- Proof regulation [serves for monitoring; to be implemented via the electronic waste proof procedure (eANV) and the central coordination office (ZKS)]. All those who are producers, owners, transporters, collectors, or disposers of hazardous declared waste are obliged to keep electronic records.
- Waste catalogue regulation (regulates the uniform designation and data management).
- Waste management company regulation (ensures the suitability and capabilities of the specialist company according to the substances to be disposed of).
- Commercial waste regulation (ensures that commercial municipal waste and certain construction-like waste are disposed of and recycled as environmentally friendly as possible).
- Landfill regulation (serves the long-term traceability of the deposit and the environmentally safe disposal of waste, for example, regarding water protection) [32].

GefStoffV: Hazardous Substances Ordinance

The Hazardous Substances Ordinance comprehensively defines measures for the regulation, classification, labelling, and packaging of hazardous substances and mixtures. In addition, it contains important regulations for the protection of employees and other persons who work with or could come into contact with hazardous substances. These protective measures include, among other things, clear guidelines for the safe handling, storage, and disposal of such substances. In addition, the ordinance restricts the manufacture and use of certain hazardous substances in order to minimise potential risks to the health of users and the environment. These comprehensive regulations aim to ensure a safe working environment, promote public security, and reduce toxic environmental impacts [33].

RoHS: Restriction of Hazardous Substances

The EU Directive 2011/65/EU restricts the use of certain hazardous substances in electrical and electronic products or devices (also known as the RoHS Directive). This directive legally regulates the burden on health and the environment caused by particularly dangerous substances. The directive aims to uniformly define the conditions for the marketing of electrical and electronic devices. As part of these conditions, the use of certain substances classified as dangerous is prohibited or restricted by limit values. This specifically includes the prohibition of mercury, lead, cadmium, hexavalent chromium, various plasticisers, and brominated flame retardants. The reduced emission of pollutants is intended to reduce the harmful effects on people and the environment and to improve the recycling of old devices. Restrictions of this kind have led to many new, less harmful technical solutions in recent years. Exceptions are only permissible if certain limit values are not exceeded, such as for lead in gas discharge lamps. Compliance with RoHS is a prerequisite for the previously described CE certification [34].

Examples of such substances used in electrical and electronic devices are listed below:

- Lead (Pb): for example, in solder and as a component in electronic devices;
- Mercury (Hg): for example, in thermometers, fluorescent lamps, and other electronic components;
- Cadmium (Cd): for example, in batteries, electrical connections, and some circuit boards;
- Hexavalent Chromium (Cr6+): corrosion-resistant material, often used in, for example, metal coatings and as a rust inhibitor in electronic devices;
- Polybrominated Biphenyls (PBB): a group of brominated chemical compounds used as flame retardants in plastics;
- Polybrominated Diphenyl Ethers (short: PBDE): flame retardants, used in plastics and textiles;
- Phthalates (e.g., DEHP, BBP, DBP, DIBP): plasticisers used in plastics, especially in cable insulation and PVC-based products.

As initially described, the laws, regulations, and directives listed here do not claim to be complete and by no means cover all areas relevant to product development. The selection made and clearly outlined here attempts to provide an exemplary representation of the legal situation. The examples given here are intended to illustrate the legal framework within which product development in general, but especially the development of more sustainable products, takes place or must take place. For the development of more sustainable products, consideration of such legal framework conditions is indispensable and needs to be expanded in many respects. Relevant laws, regulations, and directives must be researched and taken into account on a product-specific basis. Official websites of the relevant federal ministries (e.g., BMUKN, BMWE, BMJV) or the Federal Environment Agency offer possibilities for this. Official pages of the European Parliament and the European Council should also be taken into account. Other laws addressing corporate responsibility include the Supply Chain Act, which obliges companies to respect human rights and environmental standards in their global supply chains and to prevent grievances. The Federal Nature Conservation Act regulates the protection of natural habitats and requires companies to adapt their business practices accordingly to avoid environmental destruction. The Energy Saving Ordinance (short: EnEV) calls on companies to design their buildings energy-efficiently in order to reduce energy consumption. In addition, there is the EU Waste Framework Directive, which regulates the disposal and recycling of waste in order to promote a sustainable circular economy. Laws, rules, and regulations of this kind primarily address the (self-)responsibility of companies.

These and many other laws lead to fixed requirements in relation to the development of sustainable products (see Sect. 7.3). From these fixed requirements, in addition to binding regulations, a multitude of new markets and the need for new supporting technologies can be derived.

3.3 Guidelines and Standards

Guidelines and standards are documents that establish requirements for products, services, or procedures. These provide clarity about their properties and, at the same time, support rationalisation and quality assurance in business, science, technology, and administration. The application of standards is generally voluntary. However, their application demonstrates correct behaviour. In connection with the Act on Liability for Defective Products, failure to comply with standards and guidelines may also be interpreted as a fault for which a company can be held liable in the event of damage. In cases of gross negligence in handling the state-of-the-art, as documented in guidelines and standards, responsible individuals can also be held personally liable.

A large number of generally accessible ISO standards (International Standardisation Organisation) or DIN standards (German Institute for Standardisation), some of which have identical content, can be viewed via the database of the German Institute for Standardisation. Particularly relevant for the development of sustainable products are, for example:

- ISO 14062 (Guidelines for integrating environmental aspects into product development)
- ISO 14067 (Greenhouse gases—Carbon footprint of products—Requirements and guidelines for quantification)
- ISO 26001 (Guidelines for social responsibility of organisations)
- ISO 59004 (Circular Economy—Vocabulary, principles and guidelines for implementation).

Generally available guidelines, often used in product development, are often documents published by the VDI (Association of German Engineers). For the development of sustainable products, in addition to the guidelines frequently cited in this book, which relate to the development process or development methods, the following are of interest:

- VDI 2243 (Recycling-oriented product development)
- VDI 4605 (Sustainability assessment)
- VDI 4803 (Methods for efficient and gentle use of resources in companies)
- CSE (Holistic sustainability certification).

As already explained in Sect. 3.2, the selection made here provides a rough overview. Again, no claim to completeness is made at this point—rather, the guidelines and standards listed are examples in the context of developing more sustainable products. As part of the overview, it is intended to demonstrate, according to a familiar scheme, that there are a large number of guidelines and standards to consider for the development of sustainable products. In specific cases, these must be researched and taken into account separately and according to thematic focus.

A company obtains information on applicable rules and guidelines for the development of sustainable products primarily through government regulations, industry-wide standards, and certifying organisations. Relevant legal regulations can be viewed through government agencies or online databases that bundle environmental and product safety regulations.

Below are some examples of such online databases, each with a brief description:

- EUR-Lex: An EU database that provides access to European laws and regulations, including environmental regulations and product safety standards;
- Global Data: A platform that provides information on global markets, regulations, and sustainable practices;
- EcoLabel Index: A database for environmental labels and certificates, listing sustainable products and services.

Industry associations and professional societies also provide up-to-date information on best practices and current trends. Increasingly, companies, consulting firms, and law firms are using these services to inform themselves about specific legal requirements and new guidelines. Also, exchanging information with suppliers and customers helps to identify requirements for the dimensions of sustainability and to integrate them into product development. Examples of industry associations are:

- Business for Social Responsibility (BSR): A global consortium of companies that promotes sustainable business practices and provides relevant information for companies;
- German Institute for Standardisation (DIN): A German association that creates standards and norms on various topics, including sustainability and product development;
- Association of German Engineers (VDI): An industry association that informs companies in the industry about relevant regulations and standards for sustainable development.

In summary, guidelines and standards also have a direct influence on product specifications and can have both innovation-promoting and restrictive effects.

4 Product Life Cycle and Circular Economy

The following chapter deals in detail with the Product Life Cycle, its associated phases, and their further significance. As an introduction, the Product Life Cycle is generally defined and presented to create a uniform understanding for the subsequent sections. This is followed by a discussion of the concept of technical inheritance and, in extension, the significance of technology cycles and product generations. Subsequently, the principle of the Circular Economy, building on the previously described foundations, will be discussed. To introduce this, Fig. 4.1 shows a classic Product Life Cycle with associated phases.

Product life cycles and technology cycles are closely linked in product development, as technological advancement significantly influences the duration and phases of a product's life cycle. The Product Life Cycle describes the various phases of a product from product creation, through the usage phase, maintenance, to the end of life and potential reuses. The technology cycle, on the other hand, refers to the introduction of new technologies and their development from the initial innovation to full market penetration and possible replacement by a new technology.

Every product undergoes a life cycle, from product planning through the phases shown in Fig. 4.1, to disposal. Strategically, the life cycle involves the development, production, distribution, disposal, and removal from the market of entire product generations with hundreds or millions of individual product versions. Sometimes, the entire influence of a technology, in varying degrees, is considered over several product generations. An assessment can then, at least retrospectively, also be carried out using the methods of Life Cycle Assessment (see Chap. 5). The following Fig. 4.2 presents the relationships between origin, growth, maturity, and age in the technological life cycle with individual, underlying product generations.

Technology cycles influence the Product Life Cycle by enabling new features and improvements that further develop the product during its lifetime or bring

R. Lachmayer et al., *Methodology for the Development of Sustainable Products*,
https://doi.org/10.1007/978-3-662-72442-2_4

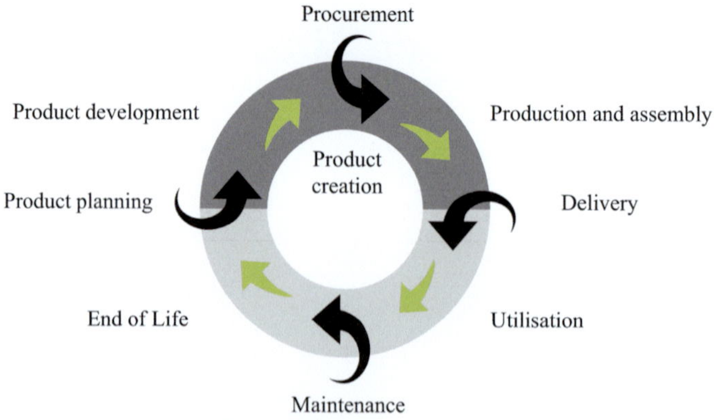

Fig. 4.1 Product life cycle with associated phases

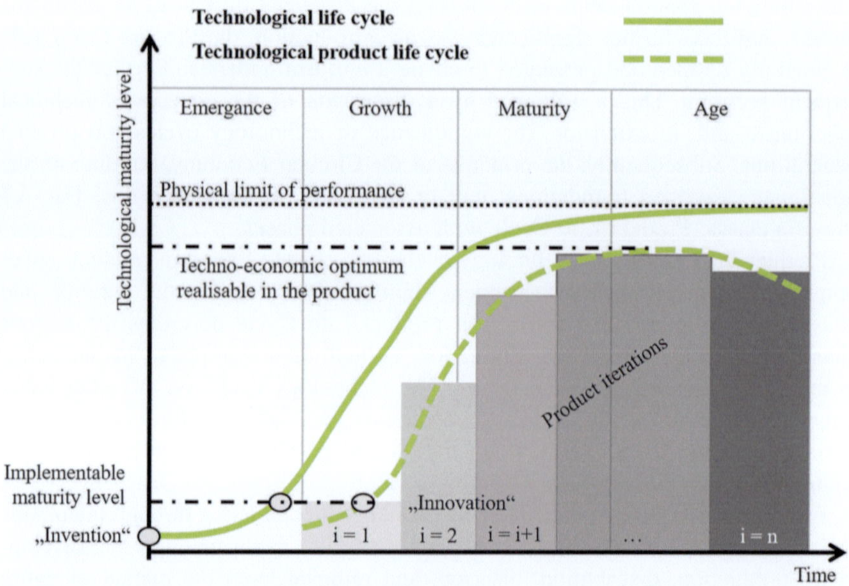

Fig. 4.2 Technology life cycle and product generations

about the need for adaptation or redevelopment. After an initially slow development of the technology, a product can be implemented and brought to market from a certain maturity level. Over product generations, the technological life cycle reaches the technical-economic optimum, with the physical limit of performance always being slightly higher and never fully exploited.

In the context of product generations, the concepts of technical evolution and technical inheritance will be discussed in more detail. The four terms—individual,

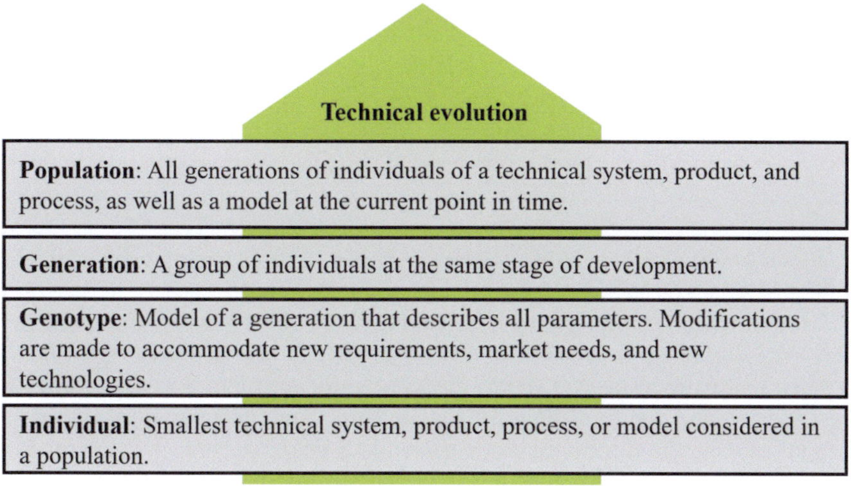

Fig. 4.3 Process diagram of technical evolution

genotype, generation, and population—reflect the constant development and collective learning in technology development. The change of perspective from one stage to the next not only promotes the optimisation of individual products but also the emergence of new paradigms and the spread of innovations in society. This principle can be best explained using a specific example—for the following explanations, the example of a VW Golf was chosen. For the development of this car, it turns out that the population includes the generations Golf 1 to Golf 8. One of these generations consists of different genotypes—in the case of the Golf, for example, an estate car or a compact car. Within the genotypes, the individual produced vehicles are seen as phenotypes, which can be characterised, among other things, in terms of tolerances and wear.

The following Fig. 4.3 summarizes the described relationships of technical evolution.

The paradigm of technical inheritance refers to the process in which technological innovations, product data, and information are transferred and developed from one generation to the next.

In product development, this means that a technical system or a product is not considered in isolation, but as part of a continuous development process in which innovations build on existing technologies. Technical inheritance also promotes standardisation and the creation of compatibility, as technologies are often developed to interoperate with older systems or products. This continuous development contributes to making new technologies more efficient, robust, and user-friendly. Product and process innovations contribute to the optimal usage of technologies and are characterised either by the market (i.e., added value for the customer) or by quality or productivity improvements. Depending on the effort, product and process innovations can be enforced in development projects. New technologies, however, cannot be enforced, even with high expenditures. These must be invented within the framework of scientific constraints. Technical inheritance and the circular economy are closely linked, as both concepts promote the continuous usage and improvement of resources. Through technical inheritance, products can, for example, be designed more modularly, which improves their repairability and reusability of individual components or assemblies, which in turn benefits the circular economy as well as quality and reliability aspects.

Based on the Product Life Cycle shown in Fig. 4.1, the concept of the circular economy builds on the idea that at the end of usage, the product is not disposed of as worthless, but that parts and materials from the product are reintegrated into the value creation process for the production of new products. Even better is when the entire product is refurbished and introduced into a second cycle—in this case, we also speak of refurbishment or reuse. The fundamental vision of the circular economy is to meet the entire demand for the production of future product generations with materials already in circulation, thereby avoiding the need for further (over-)exploitation of the Earth's natural resources. In reality, the recycling management law described in Sect. 3.2 has achieved the goal that many recyclable materials can already be recycled and reintroduced into their respective cycles today. However, as with the reuse of steel or aluminium scrap, the melting process in the furnace often requires high amounts that currently cannot yet be provided from renewable sources. Another limitation of this approach can be observed in the reuse of thermoplastics, where the original material quality is often not achieved. In this case, we speak of so-called downcycling.

Product Circularity

The degree of a product's circularity is determined by various influencing factors in the product development process. The following illustration shows some of these factors. The selected influences do not represent the entirety—they rather present a selection of relevant factors.

As shown in Fig. 4.4, factors influencing a product's circularity include, for example, the reuse of recycled materials in the design or, even better, the reuse of entire components or subsystems. For production, flexibility for combination and economical use of energy are desirable. In usage, ease of repair and low maintenance requirements are necessary. For the return, initial disassembly and sorting capabilities, as well as the required interconnected process steps, are important.

Before dealing with these relationships in detail, a look back at the three essential dimensions of technology is necessary. As already addressed in previous chapters, a technical system and the associated cycle are always characterised by all three of these general dimensions, namely material, energy, and information. In this context, humans have primarily transferred material flows, then energy flows, and in recent decades also information flows to technical systems [35].

Depending on whether the technical system is a facility for material conversion, energy conversion, or information conversion, it is often colloquially—and not entirely correctly—referred to as apparatus, machines, or devices [36].

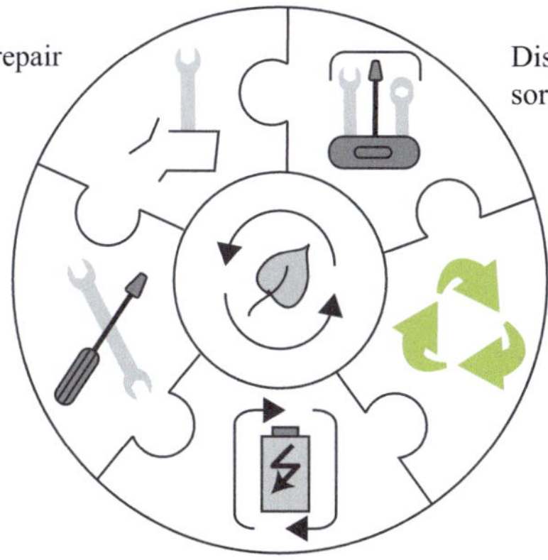

Fig. 4.4 Factors influencing the degree of a product's circularity

Table 4.1 Manifestations of the general dimensions

Dimension	Material		Energy		Information	
Measure	Flow [kg/s]	Quantity [kg]	Flow [kg/s]	Quantity [J]	Flow [bit/s]	Quantity [bit]
Characteristics	• Type of substance • Condition (e.g., pressure, temperature) • State of aggregation • Shape		• Type of energy • Distribution to components • Material carriers		• Type of information (image, language, sound) • Codes • Type of signal	

The general dimensions of material, energy, and information are well-suited for describing and analysing technical systems and their mode of operation at an abstract level. To describe systems in space and time more closely (e.g., also in the form of circuit diagrams), the dimensions of material, energy, and information also include executable operations. These operations include converting, shaping, storing, conducting, and linking, with many different technical implementations. The characteristics of the general dimensions can also vary greatly—see some examples in Table 4.1.

The following will analyse a technical system using an example from electromobility. The aim is the exemplary application of the general dimensions and operations, taking into account physical units for the description of an E-Mobility Charging System. This example serves to illustrate previously introduced theoretical concepts. The charging system under consideration includes various physical components. These include, among others, the charging cable (length in metres [m]), energy storage devices such as vehicle batteries (mass in kilograms [kg]), electrical storage elements such as capacitors (capacity in Farad [F]), and the electric vehicle itself (mass in kilograms [kg]). The material flow occurring within the system describes the transport or consumption of material resources during the charging process. The energy supplied to the system is in the form of electrical energy and is measured in watt-hours [Wh] or kilowatt-hours [kWh]. The energy flow during the charging process can be characterised by the electrical power in kilowatts [kW], whereas the energy consumption over a certain period of time is expressed by the consumed energy quantity in kilowatt-hours [kWh].

The system efficiency can be described as the ratio of the electrical energy delivered to the vehicle to the energy taken from the power grid. In addition to material and energy quantities, the information flow also plays a central role in the charging system. This particularly concerns the control and monitoring of the charging process. Relevant information quantities include, for example, the state of charge of the vehicle battery (in percent [%]) or the current charging power (in kilowatts [kW]). Information processing is done in digital units such as bits [bit] or bytes [B].

For communication between the vehicle and the charging station, standardised protocols are used, which exchange data at certain transmission rates (in bits per second [bps] or kilobits per second [kbps]). These communication processes enable the continuous adjustment of the charging process based on current system data. The energy flow typically occurs through the conversion of alternating current (AC) into direct current (DC), which is fed into the vehicle battery via the charging cable. The parallel information flow serves to monitor and control the charging process, where corresponding data is digitally encoded, transmitted, and, among other things, stored on energy cards for billing purposes.

Overall, the charging system exemplifies the networking of material, energy, and information flows within a technical system. The description is done using physical units and technical quantities, as is common in engineering system descriptions, and is systematised using general quantities and operations.

4.1 Material Cycles

Living in a real and material world necessitates those raw materials, from which materials are produced to be used in products, that are necessary for all everyday objects. This is particularly true for those of the digital economy (mobile phones, computers, laptops, printers, screens, virtual reality glasses, data centres, etc.). The goal of material flow management or the circular economy is to achieve as intensive usage of the substances extracted from nature as possible, to save resources and avoid waste. The UN Enquete Commission on "Protection of Man and the Environment" formulated the following rules in this regard as early as 1994:

- **Regeneration**: The depletion rate of renewable resources should not exceed their regeneration rate;
- **Substitution**: Usage of non-renewable energy only with equivalent replacement or higher productivity of renewable resources;

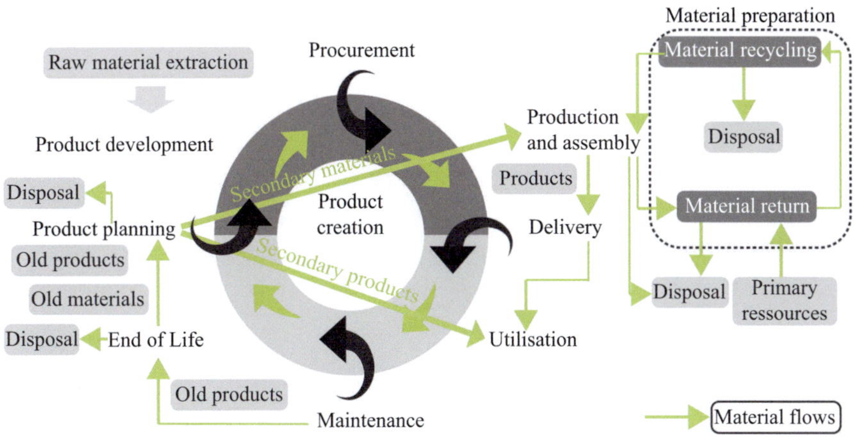

Fig. 4.5 Material return flows within the circular economy in the product life cycle; own representation based on Herrmann [38]

- **Adaptability**: Substance inputs into the environment should be oriented towards the resilience of environmental media;
- **Responsiveness**: The time scale of anthropogenic (human-made) inputs into the environment must be in balanced relation to the time scale of processes relevant to the environment's responsiveness;
- **Precaution**: Hazards and unacceptable risks to human health from anthropogenic influences are to be avoided [37].

In the following Fig. 4.5, the Product Life Cycle (see Fig. 4.1) is repeated and further supplemented with potential material flows for reuse in the sense of a circular approach.

Important terms in the figure are linked to corresponding, arrow-represented assignments. The following are some of these terms explained for better understanding:

- Primary resources: The initial introduction of raw materials into the cycle;
- Waste: Material to be disposed of as worthless;
- Recycling: The reintroduction into the raw material cycle;
- Secondary materials: The utilisation of secondary materials and their reintroduction into production;
- Thermal Recycling: The incineration of waste materials for energy recovery;
- Secondary Products: Products that have already been phased out of use once and are reintroduced as a whole or in modules for further or renewed use;
- Refurbishment: The reconditioning and value enhancement of old products to reintroduce them into a usage phase;
- Reuse: The reuse of complete products;

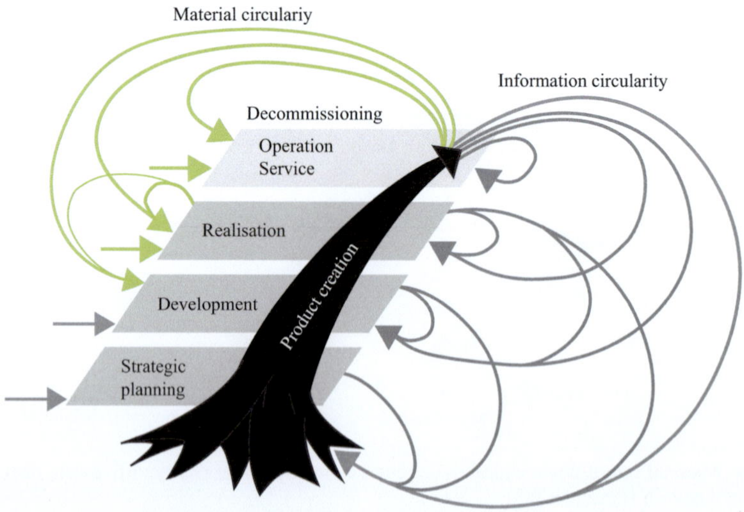

Fig. 4.6 Hierarchy of reintroduction in the circular economy; own representation based on Gräßler and Pottebaum [39]

- Upgrade: The enhancement or adaptation of existing products to changed boundary conditions.

The following Fig. 4.6 aims to extend the previously described characteristics of the circular economy by the influencing factor of a hierarchical level of consideration. The figure shows a hierarchy of the benefits of reintroduction in the circular economy. The greater the number of higher-value parts, i.e., whole products or modules thereof, that are reused, the shorter the material cycle is closed and the fewer additional resources are needed. Ideally, whole products are introduced into a second life cycle.

In addition to the examination of material reintroductions in the form of material circularity, primarily from product creation, back into operation and realisation, the approach of Gräßler and Pottebaum [39] also examines information circularity. Here, information is fed back into all phases of product creation. This assumption increases the relevance and necessity of consistent data management within organisations. The remainder of this section will focus on the details of material flow management. Material flow analysis is suitable for examining and describing material flows within a system. Such material flow analyses can be quantified and graphically illustrated using the relationships shown in Fig. 4.7.

The simplified system model consists of the elements: new device production, primary market, remanufacturing system, secondary market, and disposal. Between these elements, initial material flows (e.g., new products enter the primary market and thus the system), return flows (e.g., used devices back into the system), and processing flows (e.g., for reuse after the primary market) occur. Both the return

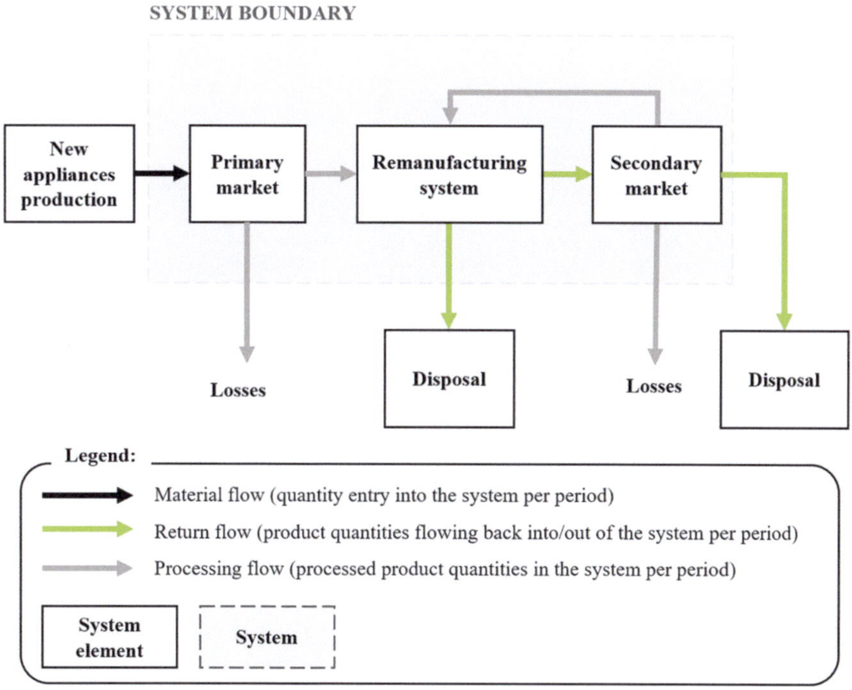

Fig. 4.7 Material flow analysis using the example of device production; own representation according to Walther et al. [40]

rate and the processing rate can be influenced by the manufacturer's decisions, such as product design or targeted take-back and distribution strategies within the business model. Products that cannot be returned leave the system via the path of losses. The processing rate, as a measure of the processed product quantities, can at least partially be influenced by active decisions of the producing company in the product development phase. Within the framework of the Circular Economy Act and the "green dot" label, it does not primarily matter at which point in the process chain a reintroduction into the cycle takes place. Also, the potential option of recycling is often not sufficient, as in many cases, neither processes nor technical possibilities for a return or recycling at the local level are available.

Analogous to the processing rate, the return rate is also significantly influenced by the manufacturer. However, this means that materials or parts are reused by the manufacturer himself for his next productions—downcycling should thus be avoided at this point. It is expected that binding return rates will be prescribed by the legislator in the coming years.

Figure 4.8 exemplarily shows the material flows per month in a producing company. In total, 200 tons are procured as material. Of this, 144 tons go into production/manufacturing. 40 tons go directly into assembly as prefabricated semi-finished products or standard parts. 20 tons are subjected to quality control and

Material flow matrix

from \ to	Receiving warehouse	Cutting	Manufacturing	Assembly	Quality inspection	Dispatch	Scrapping	Sum
Receiving warehouse		200						200
Cutting			144	40	20			202
Manufacturing		4		104	32		12	152
Assembly			8		130		6	144
Quality inspection						182		182
Dispatch								
Scrapping								
Sum	0	204	152	144	182	182	18	

Units in [t/month]

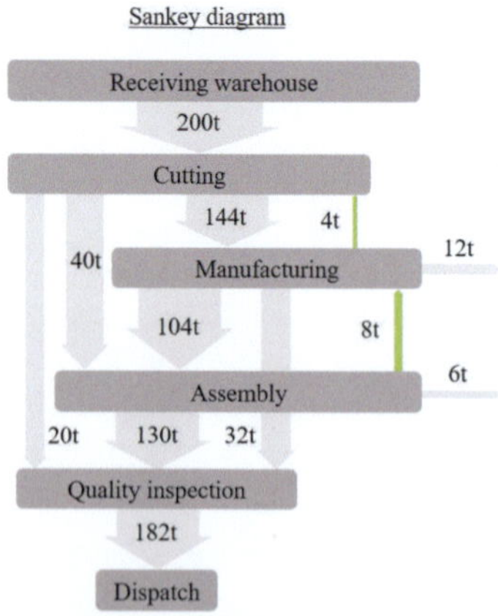

Fig. 4.8 Material flow matrix and representation of flows as a Sankey diagram

supplement the product's readiness for dispatch. In this example, 12 tons are lost in machining and 6 tons in assembly, e.g., as surplus or packaging.

Both representations in Fig. 4.9 serve to analyse material flows, but in different ways and with differing decision support. While Sankey diagrams visually and intuitively represent the flow, the material flow matrix provides detailed numerical data and relationships between the individual flow points. Sankey diagrams can be derived from material flow matrices to present the same data in an understandable and visually appealing way. In practice, material flow matrices are often used to collect and analyse data, which is then transferred into a Sankey diagram to visually illustrate the results of the analysis. Based on the consideration of approaches to material flow management and the visualisation of material flows, dealing with waste and emissions is essential. In product development, current developments are increasingly focusing on environmentally friendly handling of waste and emissions. An example of this is the use of recycled materials in the automotive industry, such as recycled aluminium alloys, which reduce material consumption and emissions. Also, the introduction of Circular economy models, such as those promoted by companies like Patagonia or IKEA, encourages the reuse of materials and reduces waste. In the electronics industry, the recycling rate of smartphones has increased through initiatives like Fairphone, which improve repairability and the recycling process.

At a national level, laws such as the previously mentioned German Circular Economy Act (KrWG) oblige companies to minimise their waste and close material cycles. These and other initiatives described in Sect. 3.2 contribute to reducing waste and emissions by integrating sustainable production methods into companies.

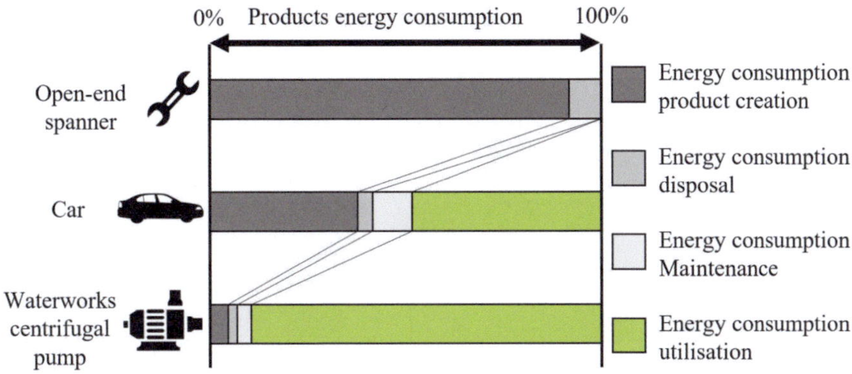

Fig. 4.9 Percentage energy consumption for the production and usage of different products based on Ehrlenspiel et al. [41]

4.2 Energy

As previously described, this chapter aims to describe the general quantities of material, energy, and information and place them in the context of the circular economy. The following paragraph refers to the quantity of energy, associated energy flows, and their impact on product development processes, forms of analysis, and the relationship to the circular economy.

In the context of the sustainability discussion, the general quantity of energy—including its management and the structure of the energy system—represents the next essential general quantity. In the context of product development, both the consumption for the production of a good and the consumption during the usage phase are optimised. However, the fundamental decision on the form of energy used in the system is also defined at the beginning of development. The fact that very different challenges arise depending on the product is shown in Figs. 4.9 and 4.10.

When considering the general quantity of energy, it can be noted that as soon as a system is supplied with energy from outside, work (unit: kilowatt-hours [kWh]) is performed. Depending on the time required for this, the power (unit: watt [W]) of a system results, which can also be stated as an energy flow.

The generation and usage of energy in the life cycle of a product vary depending on the type of product and its functionality. While a spanner requires only a small amount of energy for its production throughout its entire life and consumes no significant energy during usage, this is different for products like a waterworks centrifugal pump. The latter causes high energy consumption in operation, as it requires energy for the pumping of water and is often in use for decades. Between these extreme examples, there are often products from the mobility sector, such as cars, which show high energy expenditure in both product creation and the usage phase.

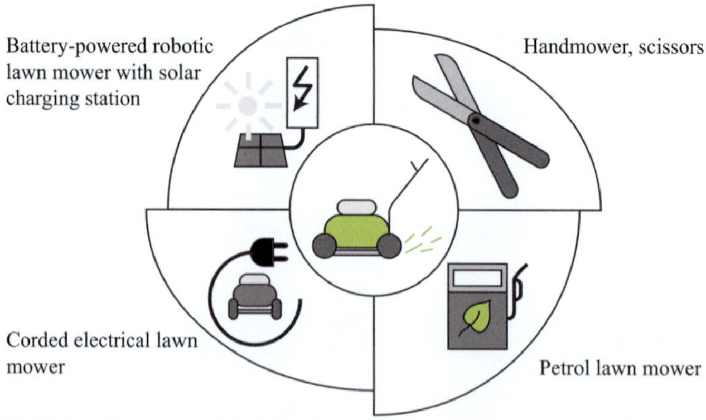

Fig. 4.10 Different types of energy supply using the example of a lawnmower

Analogous to the economic consideration of investment costs (short: CapEx) and operating costs (short: OpEx), it is equally crucial in product development to consider both the energy expenditure in production and the operational energy consumption to optimise the entire life cycle of a product in terms of energy efficiency. The following example will therefore deal more closely with different types of energy supply. For this purpose, the product example of a lawnmower is considered and executed in the context of various types of energy supply.

There are various options for energy supply for lawnmowers, each of which places different demands on product development. Battery electric lawnmowers offer the advantage of mobile usage without a cable, but their development requires powerful and durable batteries as well as efficient charging cycles to ensure an adequate runtime. Electric lawnmowers with a cable are cheaper, more efficient, and provide an uninterrupted power supply, but cable management must be considered in development to allow flexible handling. Petrol lawnmowers offer high performance and flexibility but require the use of combustion engines and fossil energy carriers, posing the challenge of minimising emissions and noise for product developers. The manual lawnmower represents the simplest form of energy supply, where no external energy source is necessary. Only the mechanical resistance and associated user-friendliness need to be optimised to reduce the effort required by users. From the perspective of product development, it is crucial to align the energy source with sustainability goals and user requirements. The choice of energy source not only affects environmental impacts but also costs, maintenance effort, complexity, and product longevity.

A key aspect of dealing with energy is the realisation that it cannot be generated by technical means. Energy, in whatever form, can only be converted from one form to another, and this conversion can be utilised. Table 4.2 shows an overview of different forms of energy, associated definitions, and some examples.

Table 4.2 Forms of energy, definitions, technical examples

Form of energy	Definition	Examples of energy carriers or storage
Chemical energy	Energy is stored in chemical bonds between atoms and molecules	Natural gas, petroleum, galvanic cell, photosynthesis, fuel cell
Electrical energy	Energy converted by the flow of electrical charges in a field	Electric motor, capacitor, accumulator
Nuclear energy	Also known as atomic energy; energy stored in bonds within the atomic nucleus	Nuclear power plant, atomic weapons, fusion generator
Kinetic energy	Energy possessed by a body due to its motion	Flying ball, flywheel, wind, moving car, running person
Potential energy	Energy possessed by a body due to its position in a force field or its configuration	Apple on a tree, reservoir, stretched spring
Radiant energy	Energy is transmitted via electromagnetic waves	Microwaves, light, infrared radiation, X-rays
Thermal energy	Energy possessed by an object due to the motion of its molecules—also referred to as heat energy	Internal combustion engine, hot water bottle, boiler, waste heat

In closed systems, the total energy remains constant; however, not all energy conversions are feasible, even if they comply with the law of conservation of energy. A conversion can only occur if the conserved quantities of the system remain unchanged before and after the conversion. For example, the conversion of kinetic energy is limited by the conservation of momentum and angular momentum. Even at the molecular level, many theoretically possible chemical reactions cannot occur because they would violate the conservation of momentum. Other conserved quantities, such as the number of baryons and leptons, limit nuclear reactions and the conversion of energy.

Electrical energy, for example, can be easily converted into many other forms of energy, such as kinetic energy through electric motors. However, with almost all conversions, a portion of the energy is lost in the form of heat, which is referred to as dissipation. Technical systems are usually not perfectly insulated, making the occurrence of energy losses inevitable. A conversion of heat into other forms of energy always requires a temperature difference and leads to an inevitable portion of dissipated energy, which is lost as heat. In technical applications, several coupled types of energy conversion often occur. A coal-fired power plant, for example, converts the chemical energy of coal into heat through combustion processes, which is then used to generate steam, which is ultimately converted into electrical power through the conversion to kinetic energy. The first law of thermodynamics, which describes energy conservation, also applies here. The second law of thermodynamics, on the other hand, stipulates that the conversion of heat into other forms of energy involves technical effort and cannot be carried out completely without losses.

In summary, it can be stated that the conversion of energy in a system is always limited by physical laws and entropy. Systems tend to reach states of maximum disorder, and a complete conversion of all forms of energy is practically impossible. Every energy conversion involves losses, and therefore, the entire system must be considered energetically. The choice of forms of energy has a direct impact on the development of more sustainable products, as it determines the energy efficiency and environmental impact of a product. For battery-electric products, such as the lawn mower described, battery technologies must be developed that have a longer lifespan, shorter charging times, and less environmental impact. Energy-efficient solutions often require the integration of intelligent control systems that optimise energy consumption. Products powered by renewable energies promote sustainable usage and reduce the resulting CO_2 emissions. From a technical point of view, it is crucial to align the energy source with the available resources and user requirements in order to create holistic, sustainable solutions in the long term and minimise the number of energy losses in a system.

Already in the nineteenth century, the engineer Sankey dealt intensively with the thermal efficiency of steam engines. It is said: “Once the need for management has been determined by balancing, it is necessary to seek ways, based on science and technological progress, to introduce cost-saving measures”. In addition, all means should be used to reduce losses and utilise waste [42]. For this purpose, Sankey calculated material and energy balances and graphically represented the steam flows, so that the strength of the arrows between the individual nodes corresponds to the quantity of the substances. These images were subsequently referred to as Sankey diagrams. Below is such a diagram in its original form (Fig. 4.11).

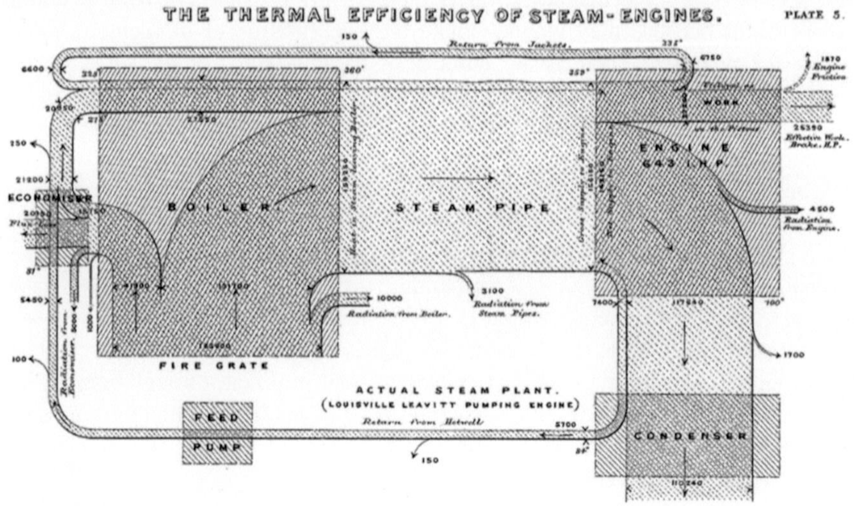

Fig. 4.11 Original representation of the thermal efficiency of a steam engine according to Sankey [43]

A Sankey diagram, according to these explanations, can be characterised as a special type of flow diagram that represents flow quantities via quantity-proportional arrows. The flows always run between two nodes or between two processes. In its early application, such a diagram illustrated how energy from a source, such as a boiler, passes through the steam process into a machine and what losses (e.g., heat) occur. These diagrams are particularly helpful for analysing and visually representing energy losses and the efficiency of machines.

Sankey diagrams are now used in the fields of energy management, process engineering, and process control. In another example, the topic of energy flow will be expanded to include an efficiency perspective. Specifically, the efficiency of vehicle headlights will be examined and considered. Generally, it can be stated that the efficiency of conventional vehicle headlights is relatively low, as a large part of the supplied electrical energy is converted into heat instead of visible light. In traditional halogen headlights, the electrical energy is mainly emitted as infrared radiation, which subsequently leads to high heat losses. This power balance shows that only a small part of the energy is used as light, while the majority is lost as heat, with efficiencies of electrical power to light for halogen systems at 3%.

In the following Fig. 4.12, the described relationships of energy conversion and efficiency are illustrated using the example of a headlight developed at the institute. The relative power of the headlight for different phases of light generation and the components in the beam path are shown.

The vehicle laser headlight system described here and shown in Fig. 4.12 has an overall efficiency of 8.7%, with the main proportion of the losses occurring in the four installed laser diodes. These losses occur because the diodes emit a total of 8.17 W or 78.3% of the electrical energy supplied to them in the form of heat. The electrical input power in the experiment shown is 10.43 W. The module emits 0.91 W of useful optical power. The values for input power, power after

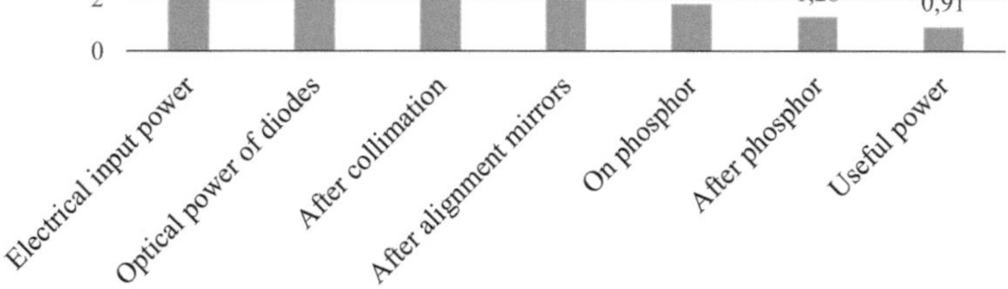

Fig. 4.12 Representation of the efficiency of a vehicle laser headlight system for phases of light generation [44]

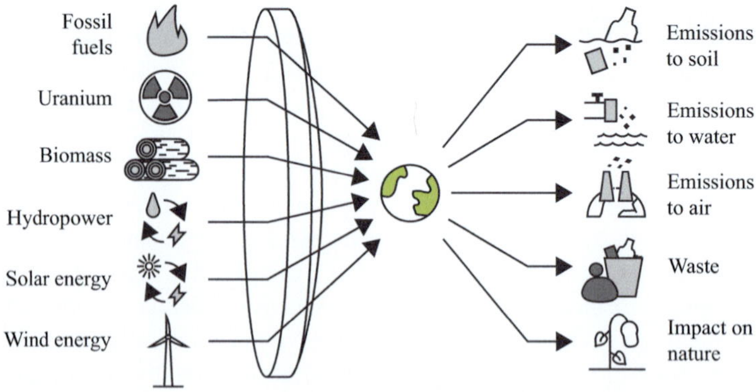

Fig. 4.13 Primary energy sources and emissions (qualitatively)

collimation, power on the phosphor, and useful power are based on conducted measurement series. All other values result from these results and take into account the separately measured efficiencies of the components in the beam path [44].

While in the classic development of technical systems, the provision of energy at the system boundary is often assumed to be given (the electricity comes from the socket, the petrol is obtained from the petrol station) and further questions in many cases are only considered under aspects of operating cost accounting, the analysis of the entire process chain plays a decisive role, especially for the ecological consideration. Of interest in such a process chain-oriented perspective are both the initial provision and the composition of the available and used primary energy.

The following Fig. 4.13 schematically shows the primary energy sources that are fundamentally available. These are differentiated into renewable and non-renewable energy sources. In addition, the types of emissions resulting from these primary energy sources are listed and also schematically represented.

The assessment of measures in terms of their sustainability or specifically their eco-efficiency, such as the conversion of road traffic to electromobility, therefore directly depends on the provision of the respective form of primary energy for electricity generation.

The following are some figures on primary energy consumption for Germany, both absolute and by energy carriers, researched and prepared for the years 1990 and 2023, and presented. For the two years, the following primary energy consumptions and corresponding energy carrier shares are obtained. The listing is made in this context in descending order from the largest to the smallest share of the energy carrier (Table 4.3).

A terawatt hour (TWh) corresponds to the enormous amount of 10^{12} Wh or 1 billion kilowatt hours (kWh). For comparison: With one kWh of electricity, you can blow-dry your hair for about an hour, prepare 70 cups of coffee, or operate a laptop for 10 h. A smartphone consumes about 4 kWh per year with frequent charging, although the associated primary energy consumption due to inefficient

Table 4.3 Primary energy consumption by energy carriers 1990 and 2023 according to BMU [45]

1990	2023
35.1% Petroleum	35.6% Petroleum
21.5% Lignite	24.8% Gases
15.5% Hard coal	19.6% Renewable energy
15.5% Gases	8.7% Hard coal
11.2% Nuclear energy	8.3% Lignite
1.2% Renewable energy	2.3% Other energy sources
	0.7% Nuclear energy
Total: 3905 TWh	Total: 2982 TWh

charging processes and losses in the network is likely to be twice as high. Comparing the two years, significant efficiency gains can be seen, which are reflected in almost triple gross domestic product and a primary energy requirement that has fallen by about 25%. There is also a significant shift from fossil to renewable primary energy forms.

4.3 Information

According to the previous sections, after discussing the general sizes of material and energy, the size of the information should now be described and put into context with product development and a desired circular economy. To begin with, the term information should be defined. When asking what information actually means and how it can be described, the attempt to explain using the so-called knowledge pyramid according to Aamodt and Nygård (1995) is a sensible approach. The knowledge pyramid is schematically represented and expanded with examples in the following Fig. 4.14.

In an initial step, data is generated from symbols and signs using syntax, which already formulates a statement in this process. The next step towards information is characterised by assigning a meaning to the previously generated data and statements. This process is also equated with the term semantics. After a meaning has been assigned to the linguistic signs, they can already be called information. The knowledge pyramid also describes one last step, the transition from information to knowledge. This happens by linking information with experiences and interpreting them. This step or transition is called pragmatics.

Digital Circular Economy

The principle of the digital circular economy describes the extension and implementation of the previously described basics on the subject of circular economy with smart, digital, sometimes intelligent, and information-specific factors. The term digital circular economy fundamentally includes the collection, integration, and analysis of data and information in the context of a circular economy. The explicit goal

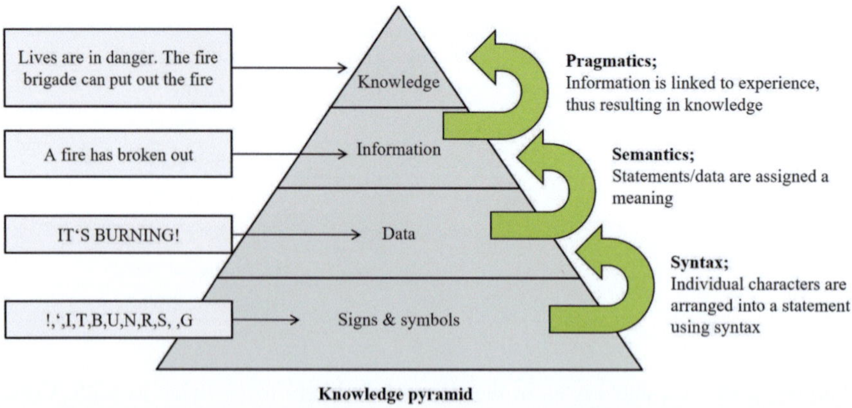

Fig. 4.14 Knowledge pyramid according to Aamodt and Nygård [46]

is to be able to make evaluations more objectively and to improve or solve the currently existing problems of end-of-life, among other things, in the recycling industry, through suitable data information systems. To efficiently store, manage, and retrieve this information, databases are used as structured systems for the long-term storage and rapid processing of data. The following list includes selected and relevant databases in the context of ecological evaluations, each extended with the respective web address.

- **ProBas**: Process-oriented basic data, library for lifecycle data
  - Categories: Energy, materials and products, transport, disposal
  - Link: probas.umweltbundesamt.de
- **GaBi**: Holistic balancing
  - Primary data categories: Agriculture, construction and design, chemicals and materials, selected consumer goods, education, electronics and information technology, energy and utilities, as well as renewable resources
  - Link: gabi.sphera.com
- **Ecoinvent**:
  - Life cycle inventory data on energy, materials, waste disposal, agricultural products and processes, transport, electronics, metal processing, and building ventilation.
  - Link: ecoinvent.org
- **ELCD**: European Platform on Life Cycle Assessment
  - Process datasets on transport services, material production, energy supply, and land use.
  - Link: eplca.jrc.ec.europa.eu

However, all these databases inadequately address the real products, their quantities, and their usage state in the field. Specific problems for an efficient circular economy include:

- Underdeveloped information availability, particularly regarding the quality and availability of recyclates.
- Increased transport and search costs, as well as increased effort for calculation and general information procurement.
- Distorted perception of potential customers about the quality of products and the inferiority of secondary materials.
- Technological effort and lack of perception of the benefit of recyclability as a competitive advantage.

As a result of these problems, fundamental solutions emerge in three directions:

1. Products should be designed so that materials can be easily separated from each other by type. This concerns disassembly, labelling, and dealing with multi-materiality and is further discussed in Chap. 8.
2. Gentelligent components can also inherently store data about their life cycle and thus provide information for their further use directly. Examples of how this is possible are shown in Fig. 4.15.
3. Gentelligent components, such as inherent sensors, directly integrate sensor functions into their structure, enabling continuous monitoring of the condition and environmental conditions. An example is laser-structured strain gauges, which measure changes in strain and tension of a component based on microsensors, particularly used in structural monitoring of machines and buildings. Fibre optic sensors can be integrated along components and enable highly precise measurement of temperature, pressure, or mechanical loads. Components dynamically adapt to temperature changes through their material properties, resulting in improved efficiency and product lifespan. These technologies contribute to the development of self-monitoring, resource-saving systems that respond early to changes and thus reduce maintenance costs and risk of failure.

Another approach is shown in Fig. 4.17. Today, so-called Cyber-Physical Systems appear in many places. These communicate, primarily for functional reasons, with the internet and other surrounding systems. Product-Service Systems (PSS) can, for example, automatically reorder consumables, initiate maintenance, and transmit statements about their usage state. A further expansion of these capabilities could enable the creation of markets by the products themselves, given the provision of sufficiently large AI-supported databases. The concept or paradigm of the Internet of Things (IoT) assumes a ubiquitous presence of a multitude of things, or artefacts, or objects in the environment. These can interact with each other and communicate and collaborate with other, also networked, things or objects through wireless or conventionally wired connections and unique schemas. This can subsequently lead to the creation of new applications or services and the achievement of common goals.

Challenges of Large Amounts of Data

Dealing with extreme amounts of data presents a number of challenges in terms of efficiency, parallelism, access control, and data security.

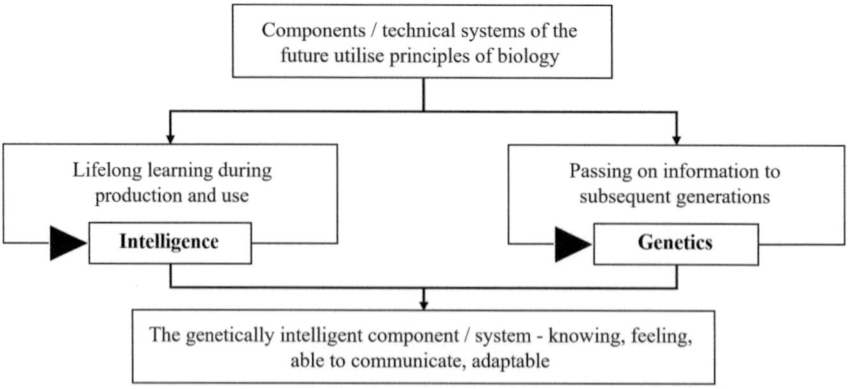

Fig. 4.15 Schematic overview of gentelligent components [47]

Existing solutions include so-called "Relational Database Management Systems" (RDBMS), which, based on a rational model, offer an intuitive and simple way to represent data and tables. RDBMS are characterised by a separation of logical data structures and physical storage structures. The "Structured Query Language" (SQL) is often used to query data, making the data applicable for any information need where different data points relate to each other.

With the help of various algorithms, artificial intelligence (e.g., multi-agent systems) can also recognise patterns and check forecasts in various, sometimes extremely complex data sets. Due to the expected very large volume of data in the digitalisation of the circular economy, there is a future need to expand capabilities.

The third aspect is the usage of data for the development of new raw materials, material cycles, and business models. The challenges here lie in determining and harmonising available, in-process, and required quantities, as well as the subsequent necessary quality determination and documentation. A central challenge in the consistent availability of data along product life cycles is that data often exist in different formats and are collected by various actors (e.g., manufacturers, suppliers, users), which complicates the integration and usage of this data. This lack of interoperability hinders effective data usage and information, which are necessary to develop new materials that meet the requirements of technical feasibility and the desired circular economy. In addition, data about the entire life cycle of a product is often incomplete or insufficiently documented, which hinders precise traceability and analysis of material flows and their reusability. To develop new business models from this information, companies must create systems that allow for seamless capture and processing of data on development, design, production, usage, and recycling. This necessarily requires close cooperation between all participants and the introduction of standardised and transparent data management solutions. Innovative solutions for material cycles and sustainable business models can only be realised with a consistent database.

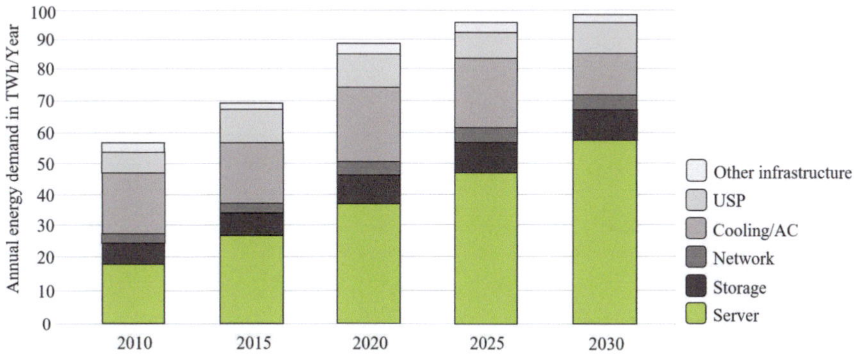

Fig. 4.16 Energy demand of data centres in Europe (forecast from 2020) according to Hintemann et al. [48]

Finally, the influence of the far-reaching digitalisation on the overall energy demand and the associated emissions should be noted. The following Fig. 4.16 shows in this context the growing electricity demand of data centres in Europe from currently just under 100 TWh per year (equivalent to 360 PJ per year).

The generation, storage, and transmission of data require significant energy expenditures. Data centres, which are responsible for processing and storing data, consume large amounts of electricity, often still from fossil sources, leading to CO_2 emissions. The network infrastructure that enables data traffic also requires energy for the operation of servers, routers, and other devices or infrastructure, which also contributes to the emission of greenhouse gases, among other things. In addition, the manufacturing and disposal processes of hardware components necessary for data processing can cause additional environmental impacts. Therefore, it is important to consider the energy consumption and emissions associated with the respective digital infrastructure to improve its overall sustainability.

An extension of this statistic deals with the previously described and ever-increasing use of artificial intelligence (AI) applications. The following comparison is intended to show that the increasing use of such AI-based technologies causes a huge increase in energy demand. Figure 4.17 shows a comparison between a conventional Google search and other AI-based search queries.

With the increasing use of artificial intelligence, it is expected that energy demand will increase significantly in the near future. This is mainly due to the ever-increasing complexity of the algorithms and the increasing amounts of data needed for the training and operation of AI systems. At the current time, it is difficult to quantify the exact extent of this increase, as the required data volumes and the associated energy requirements have not yet been fully recorded.

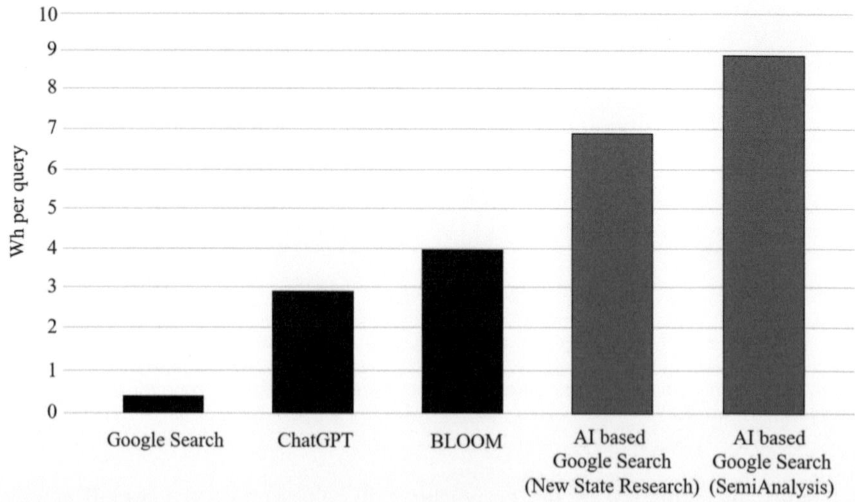

Fig. 4.17 Estimated energy consumption for various search systems per request according to de Vries [49]

5 Footprint and Assessment

The assessment of products in terms of their ecological, economic, and/or social sustainability presents companies, researchers, and policymakers with complicated and increasingly complex challenges. A central problem is that there is no universal definition or method to measure sustainability in a universally valid and holistic form. Different industries and interest groups prioritise different sustainability goals, leading to a multitude of sometimes difficult to compare or opaque assessment approaches.

This chapter examines various approaches and methods for assessing the one-dimensional and multi-dimensional sustainability of products. It aims to demonstrate how corporate responsibility and innovative assessment approaches contribute to illuminating the topic of "sustainability" in all its facets. The goal is to define a framework that helps users in practice make informed decisions by considering not only the economic viability of a product, but also its impact on the surrounding environment. Particular emphasis is placed on holistic models that enable the impact of a product throughout its entire life cycle to be analysed.

In addition to the theoretical foundation, practical examples and case studies will be presented that illustrate successful applications of these assessment methods. By engaging with proven practices and innovative approaches, it is hoped to contribute to further advancing sustainable developments in product design. In a world where the demand for sustainability is becoming ever louder, a comprehensive and holistic assessment of products can be the key to responsible and sustainable decisions.

The methods and models described in this chapter can all be categorised, as follows, into the integrating sustainability triangle depicted in Fig. 5.1.

Depending on the characteristics of the respective methods, an assignment within the sustainability triangle is made. This allows for the identification of approaches that focus solely on one dimension of the sustainability triangle, as

R. Lachmayer et al., *Methodology for the Development of Sustainable Products*,
https://doi.org/10.1007/978-3-662-72442-2_5

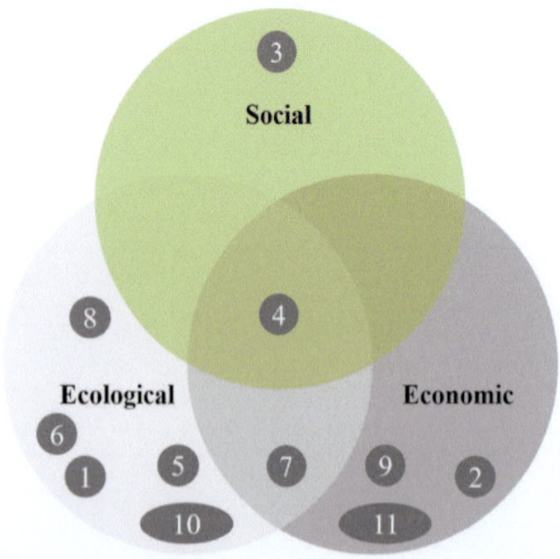

Nr.	Assessment tools
1	Life Cycle Assessment
2	Life Cycle Costing
3	Social Life Cycle Assessment
4	Sustainable Life Cycle Assessment
5	Carbon Footprint
6	Ecological Footprint
7	Material flow analysis
8	Material input per service
9	Eco efficiency analysis
10	Method of ecological shortage
11	Green Accounting

Fig. 5.1 Evaluation tools in the integrating sustainability triangle

well as other approaches that take into account the tension of multiple dimensions in the assessment.

Relevant for the assessment of product sustainability is the capture of dimension- and process-specific indicators as well as elementary flows. These can be distinguished, among other things, into resources (in energetic, material, and personnel form), emissions into air, soil, and water, and security (health, environment, and other needs).

The following will first introduce and explain one-dimensional evaluation methods.

5.1 One-Dimensional Evaluation Methods

For one-dimensional assessment and the associated quantification of sustainability, three frequently used evaluation methodologies for calculating so-called "footprints" can be identified:

- Product Carbon Footprint
- Ecological Footprint
- Water Footprint.

These different versions of footprints all aim to make the impact of a product or process quantifiable and comparable. However, each of these footprints differs in the unit and the measured object.

The **Product Carbon Footprint (short: PCF)**, for example, allows for the capture of the amount of greenhouse gases that a person emits over a certain period

Table 5.1 Factors for calculating greenhouse gas potentials according to IPCC [50]

Greenhouse gas	Greenhouse gas potential in CO_2 eq. according to IPCC07	Important sources
Carbon dioxide (CO_2)	1	Fossil fuels
Methane (CH_4)	25	Cattle farming, biomass usage
Nitrous oxide (N_2O)	298	Fertilisers, fossil fuels
Fully halogenated hydrocarbons (HFCs)	7.390–12.200	Aluminium production, semiconductor production, fire safety, extinguishing agents
Partially halogenated fluorocarbons (HFCs)	124–14.800	Coolants, propellant (e.g., in sprays)
Sulphur hexafluoride (SF_6)	22.800	Electrical industry, insulation gas (e.g., for soundproof windows)
Nitrogen trifluoride (NF_3)	17.200	Production of flat screens and solar cells

of time, a product during production or a process causes. As this form of assessment is an output-oriented view, any emissions of greenhouse gases over a certain period or within a system boundary are considered and balanced. The balance sum of all caused greenhouse gases is given in the total amount of CO_2 equivalents (short: kg CO_2 eq.), which requires a conversion of the different greenhouse gases. According to the Intergovernmental Panel on Climate Change (short: IPCC), the greenhouse gas potential of various greenhouse gases can be stated as follows (Table 5.1).

Thus, the Product Carbon Footprint is a life cycle assessment that focuses "merely" on one impact category—in this case, the "climate change" category. The guidelines for conducting and thus actually applying this method can be found in EN ISO 14067 ("Greenhouse gases—Carbon Footprint of products") and underline the need for consistent, transparent, and scientifically high-quality documentation of the respective assessments.

A specific example of a cooling system in the food industry would be an industrial cooling system that uses R-134a (1,1,1,2-Tetrafluoroethane) as a refrigerant, which has a greenhouse gas potential of about 1.430 CO_2 equivalents. These cooling systems are typically used for the storage and cooling of food in large storage rooms and in food processing.

An example of accounting for the Product Carbon Footprint of cattle with a focus on methane emissions is the consideration of cattle farming in agriculture, especially meat production. Cattle produce methane (CH_4) as a by-product of the fermentation of food in their rumen, a process initiated by microbes in the digestive system. This methane is largely released into the atmosphere through belching.

The emissions of methane by cattle represent a significant part of the CO_2 equivalent of the Product Carbon Footprint. According to [51], 100 kg of methane per year per animal can be extrapolated in intensive production systems. Specifically, for the product in the form of 100 g of beef (about 30% of a cow is meat), an average of 50 kg CO_2 eq. or a land use of 175 m^3 per year can be calculated [52].

Furthermore, the approach of the **Ecological Footprint** offers another way of capturing the biologically productive land area needed for a specific process in a holistic form [53]. The result of this assessment is given in "hectares of land area" (For comparison: One hectare equals 10,000 m^2). This choice of unit can be explained by the definition of the Ecological Footprint as the "consumption of biocapacity" for a product or process. This total consumption of biocapacity is made up of the sum of the consumption of biocapacity within production processes within a considered system boundary, or the products contained therein, and the difference in biocapacity input and output into the system. The biocapacity itself can be divided into the following five categories:

- Built-Up Land: Built-up or used land
- Forest Land: Forest areas
- Fishing Ground: Fished area
- Grazing Land: Grazing area
- Cropland: Harvesting areas.

The quality of areas used for different purposes, such as agriculture, forestry, or grazing, is determined by many factors. These factors influence both the yield capacity and the ecological sustainability of the respective areas. The following Table 5.2 summarises the main criteria to be considered in the assessment of land quality. They include both natural conditions, such as soil type and quality, water availability, and climate, as well as anthropogenic influences, such as usage history and maintenance practices. A thorough assessment of these criteria is essential to ensure sustainable management and long-term usage of the areas (see Table 5.2).

In Germany, the usage of the total area of 35.8 million hectares is distributed as follows:

- Agricultural areas: 50.4% (18.0 million ha)
- Forest areas: 29.9% (10.7 million ha)
- Settlement and traffic areas: 14.6% (5.2 million ha)
- Other: 5.1% (1.8 million ha).

The Regional Atlas of the Federal and State Statistical Offices [54] can display nationwide land use. The daily land consumption in Germany currently amounts to approximately 50 ha. This land consumption primarily affects agriculture and forestry.

The Water Footprint can be presented as a third example of one-dimensional evaluation methods. It captures the direct and indirect water consumption that occurs during a process, such as the realization of a product. This input-oriented view allows for a comprehensive collection of all resource inputs into the system

Table 5.2 Collection of criteria for describing land quality

Criterion	Description
Soil quality	• Soil type (sandy, loamy, clayey) • Soil fertility (nutrient content) • Soil structure (aeration and root growth) • pH value (nutrient availability)
Water availability	• Groundwater level (availability for plants) • Precipitation conditions (regional rainfall amounts)
Topography	• Slope inclination (suitable for agriculture and pastures) • Erosion (soil susceptibility)
Usage history	• Previous usage (overuse, soil fatigue) • Nature reserves (biodiversity and ecological quality)
Soil management and care	• Crop rotation (sustainability in agriculture) • Grazing (pasture maintenance, sustainable usage)
Climate	• Frost risk (impact on cultivation) • Seasons (length of the growing period)

in the form of water. Within this data collection, different forms of the resource are distinguished, such as precipitation, surface, sea, brackish, groundwater, and fossil water.

5.2 Multicriteria Decision-Making Methods

In the tension field of multidimensional sustainability consideration and assessment, understanding technology as a “means to an end” offers engineers the opportunity to pursue sustainability goals. By using technologies as “communication in engineering, natural, social, and economic disciplines” as well as “artefacts of (current) developments”, the use of innovative technologies forms the basis for sustainability-oriented decision-making [55].

According to Lindemann [56], the decision-making process can be divided into conscious and unconscious decisions. Conscious decisions do not have to be purely rational, but can also result from purely intuitive, improvised, or methodically supported decision-making. Regardless of this form of decision-making, the goal is to create as objective a basis for decision as possible to minimize the risk of wrong decisions. From the perspective of product development, such incorrect decisions result in significant economic and ecological impacts throughout the entire product life cycle. The automotive industry, for example, shows that these consequences can be both cost-intensive and far-reaching for organisations. For instance, recall actions of millions of vehicles lead to high organisational and financial expenditures for the producing company.

Preselection	Scoring	Influence matrix	Weighting
Assessment based on important, clear, and easily assessable criteria (fixed and range requirements, feasibility)	Assigning points for the degree to which each criterion is fulfilled:→ The total value is the sum of the points for all criteria for a given alternative.	Breaking down dependencies helps identify conflicting goals.	Assessment and prioritization of alternatives based on the relevance of criteria (linear/ progressive)

Fig. 5.2 Process chain of methodical decision-making through point evaluation based on VDI 2225 [57]

The general process of methodical decision-making with various alternatives is shown in Fig. 5.2—starting with the preselection and determination of clear criteria. Subsequently, as part of an initial evaluation, the degree of fulfilment of the individual criteria is depicted in points. The sum of the points of all criteria of an alternative forms the total value of this alternative. In a subsequently constructed influence matrix, dependencies between the criteria are broken down to identify specific goal conflicts. A final weighting of the different criteria allows an evaluation and prioritisation of the alternatives depending on the relevance of the criteria (linear or progressive).

This basic process chain can be found in further tools reflected in decision-making. Similar steps can be identified in both the utility analysis and the technical-economic analysis (see VDI 2225) [57]. Table 5.3 presents the various aspects according to Lindemann [56] for the utility analysis according to VDI 2808 [58].

Regardless of the form of these tools, further aspects can be added in the form of additional ecological or social criteria, so that sustainability-oriented decisions can be made using conventional tools.

The pairwise comparison is an evaluation method in which alternatives or criteria are compared with each other to determine a ranking or weighting [59]. The process involves systematically evaluating each alternative or criterion against each other to determine the relative importance or preference. Typically, a scale with numerical values from 1 to 9 is used to quantify the strength of preference or superiority of one alternative over another. These evaluations are then used to create an aggregated overall ranking of the alternatives. In the case of the *Analytic Hierarchy Process* (AHP), the pairwise comparison is a central method. Here, the decision-makers are asked to compare each alternative concerning a specific criterion and assess how much stronger or weaker one option is compared to the other. These evaluations are entered into a pairwise comparison matrix, which is analysed to determine the final weightings and rankings of the alternatives. An example of a scale in pairwise comparison is:

Table 5.3 Comparison of various tools of methodical decision-making according to Lindemann [56]

Sequence	Sub-step	Utility analysis	Technical-economic analysis
1	Identification of goals or evaluation criteria that must be used for solution variants, using the requirement list and a guideline	Establishment of a goal system (goal hierarchy) graded according to dependency and complexity, based on the requirement list and other general conditions	Compilation of important technical properties, as well as the desires and minimum requirements of the requirement list
2	Examination of the evaluation criteria regarding their importance for the overall value of the solution. Possibly defining weighting factors	Gradual weighting of the goal criteria (evaluation criteria) and possibly eliminating insignificant criteria	Determination of weighting factors only in the case of significantly different importance of the assessment of the evaluation criteria
3	Compilation of the property sizes applicable to the individual solution variants	Establishment of a goal size matrix	Not generally provided
4	Evaluation of the property sizes according to value perception (0–10 or 0–4 points)	Establishment of a goal value matrix using a point evaluation or value functions; 0–10	Point evaluation of the properties: 0–4 points
5	Determination of the total value of the individual solution variants, usually with reference to an ideal solution (value)	Establishment of a utility matrix taking into account weights; determination of total utility values by summing up	Determination of a technical value by summing up without or with consideration of weights, with reference to an ideal solution, possibly the determination of an economic value based on manufacturing costs
6	Comparison of the solution variants	Comparison of the total utility values	Comparison of the technical and economic values; Establishment of an s-(strength)-diagram
7	Estimation of evaluation uncertainties	Estimation of goal size controls and utility distributions	Not explicitly provided
8	Search for weaknesses to improve selected variants	Establishment of utility profiles	Identification of properties with low scores

Table 5.4 Exemplary pairwise comparison of light sources for the assessment of ecological/economic sustainability

Criterion	Halogen versus LED	LED versus discharge	Halogen versus discharge
Energy efficiency	LED: 9	LED: 5	Discharge: 5
Lifespan	LED: 9	LED: 3	Discharge: 7
Disposal/environmental impact	Halogen: 5	LED: 5	Halogen: 3
Light quality	LED: 3	LED: 1	Discharge: 3
Purchase cost	Halogen: 3	LED: 3	Halogen: 5
Heat generation	LED: 5	LED: 5	Discharge: 5

- 1: Both alternatives are equal.
- 3: One alternative is slightly preferred.
- 5: One alternative is significantly preferred.
- 7: One alternative is very significantly preferred.
- 9: One alternative is extremely preferred.

Pairwise comparison allows for precise and transparent decision-making by structuring the subjective perceptions of decision-makers and translating them into a quantitative assessment.

As a simple example of a pairwise comparison, light sources (halogen lamps, LED lamps, and discharge lamps) are evaluated based on criteria using points. The criteria listed in Table 5.4 address both the ecological and economic dimensions of sustainability.

The results of the point evaluation show that LED lamps are the superior alternative. They combine the best values in key sustainability aspects such as energy efficiency and lifespan, giving them a dominant position in the pairwise comparison. Discharge lamps are a better choice compared to halogen lamps, particularly in terms of energy efficiency and lifespan.

Alternatively, the *Multi-Criteria Decision Analysis (MCDA)* includes a variety of methods for simultaneously assessing multiple alternatives based on various criteria. Unlike the AHP, MCDA can consider both numerical and qualitative evaluation scales and is more flexible in its application. Commonly used MCDA methods include the Weighted Sum Model (WSM), where alternatives are evaluated based on weighted criteria. The SWOT analysis (Strengths, Weaknesses, Opportunities, Threats) is a strategic planning method that helps companies evaluate various action alternatives at different abstraction levels early on. It is often used in the early stages of decision-making to consider both internal and external factors. By applying the analysis, decision-makers can develop and select the best possible strategies and minimise risks.

5.3 Multidimensional Evaluation Methods

In contrast to one-dimensional evaluation methods, such as the *Product Carbon Footprint (PCF)* or *Life Cycle Costing (LCC),* which each address the ecological or economic dimension of sustainability, multi-dimensional evaluation methods focus on the tension between absolute dimensions. Drawing on the integrative sustainability triangle, the need for socio-ecological or eco-efficient approaches can be derived.

Two significant approaches to evaluate the sustainability of products and processes in a multi-dimensional and as holistic as possible way are the *Material Input per Service Unit (MIPS)* and the *Life Cycle Sustainability Assessment (LCSA).* These approaches enable a comprehensive analysis of a product's impacts along its Product Life Cycle. While MIPS focuses on the resource intensity and efficiency of a product during its lifespan, LCSA provides a systematic framework for assessing the overall ecological, economic, and social sustainability. By applying these approaches, companies can make informed decisions about product development and contribute to promoting more sustainable consumption patterns.

As "Material Input per Service Unit" (MIPS), the Wuppertal Institute for Climate, Environment, Energy [60] has developed a method for estimating resource consumption and the resulting environmental impacts of a product. For this estimation, it is assumed that every product (also in the sense of a non-physical product or a product with a service component) carries an "ecological backpack" that will be filled with inputs from various categories over the course of the product's life:

- **Biotic raw materials** (Plant biomass from managed and unmanaged areas)
- **Abiotic raw materials** (Mineral resources, fossil fuels)
- **Soil movements in agriculture and forestry** (Mechanical soil cultivation, erosion)
- **Water** (Surface, ground, and deep groundwater)
- **Air** (Chemical transformations, combustion, physical changes).

The corresponding data are collected taking into account product-specific process chains. Crucial for a targeted data collection at this point is the system boundary between the technosphere (human activities) and the surrounding ecosphere (natural environment). The exchange between these spheres forms the balance sum [60]. To gain a more comprehensive understanding of MIPS, the procedure for implementing the method will be discussed further. An initial point of reference is the process chain of the MIPS method. The following list enumerates the associated phases according to Schmidt-Bleek:

1. Definition of goal, object, and service unit
2. Representation of the considered Product Life Cycle as a process chain
3. Data collection of inputs and outputs of each process and representation as a process image
4. Calculation of material inputs from "cradle to product"

5. Calculation of material inputs from "cradle to grave" including upstream and downstream processes
6. Calculation of material inputs per service unit (MIPS)
7. Interpretation of the results.

Starting from the definitions of the goal, the considered object, and the considered service unit, the associated Product Life Cycle is represented as a process. For the respective process steps, inputs and outputs are defined and represented as a complete process image in step 3 as a result of data collection. In the subsequent steps 4 and 5, the material inputs are first calculated from the "cradle to the finished product" and the "cradle to the grave" including all processes upstream and downstream of production. These material inputs are then related to the achieved service units, so that an interpretation of the results allows a final assessment of the products.

To explain this method from the perspective of the users, the Wuppertal Institute has created various guides and templates for simplified data collection and published them as open-source information. A detailed example (including the production of pig iron) of the practical application of the MIPS method can be found in Ritthoff et al. [60].

Building on the idea of the MIPS method that each product fills an individual backpack with ecological impacts, the approach of the "Life Cycle Sustainability Assessment" (LCSA) extends this idea to include the additional dimensions of economy and social aspects. As shown in Fig. 5.3, the LCSA consists of the ecologically oriented Life Cycle Assessment (LCA), the economic Life Cycle Costing (LCC), and the Social Life Cycle Assessment (SLCA).

Various databases can be used to conduct Life Cycle Sustainability Assessments (short: LCSA) and Material Input Per Service Unit (MIPS), providing specific environmental data and information on different product life cycles. A well-known database is Ecoinvent, which offers extensive data and information on emissions, energy consumption, and raw material usage across various industries. GaBi is another widely used database that provides detailed life cycle data and

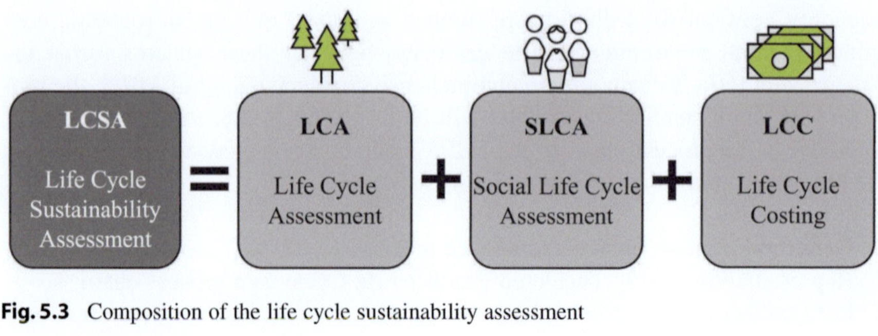

Fig. 5.3 Composition of the life cycle sustainability assessment

scenarios for assessing environmental impacts, energy consumption, and material flows. OpenLCA is an open-source database that offers a platform for life cycle assessment analysis with a variety of data sources and methods. The European Platform on LCA (short: EPLCA, link: https://eplca.jrc.ec.europa.eu/ELCD3/) provides detailed, EU-specific life cycle data aimed at the holistic sustainability assessment of products in various sectors. Various databases facilitate the consistent assessment of environmental impacts and contribute to the development of more sustainable products and processes.

This methodology aims to assist organisations in identifying ecological, economic, or social weaknesses or hotspots within the product life cycle through various methodological analyses. The economic Life Cycle Costing (LCC) and the ecological Life Cycle Assessment (LCA) will be explained in more detail below.

The approach of Life Cycle Costing (short: LCC) aims to determine the costs of a product over one or more product life cycles [61]. The results of the LCC can be used to support decision-making in the development of new products or product generations. The analysis and identification of cost-intensive life cycle phases and the processes contained therein enable a comparison of different product variants and associated strategies. Regardless of the approach used, the general procedure of the LCC can be divided into four phases:

1. Definition of objectives and scope;
2. Development of an LCC model
3. Calculation of costs over the product life cycle
4. Interpretation and review of the results.

Within the models, two basic approaches can be distinguished. These are the analogy and parameter models. While analogy models examine corresponding costs based on historical data and similar systems, parametric models represent costs based on algorithms, variables, and parameters.

In addition to this purely economic examination of the product life cycle, the question also arises as to how and in what form an analysis of the ecological impacts is possible.

The concept of Life Cycle Assessments (LCAs) or eco-balancing is described in the ISO guideline 14040 ff. [62] as the "recording and systematic analysis of the ecological impacts of products throughout their entire life cycle" ("from cradle to grave"). Consequently, all processes in production and those upstream and downstream are counted as ecological impacts. In summary, an LCA is the complete recording and assessment of the environmental impacts of a product over its entire life. For this assessment, certain resource flows, such as raw materials, energy, or resulting emissions, are quantitatively recorded as process steps and outputs. In practice, LCAs often serve to support ecologically oriented decision-making processes within organisations.

The execution of an LCA generally takes place in the form of the four phases shown (see Fig. 5.4). Starting with a definition of the objective and the examination framework under consideration, the target group, the functional unit, the product

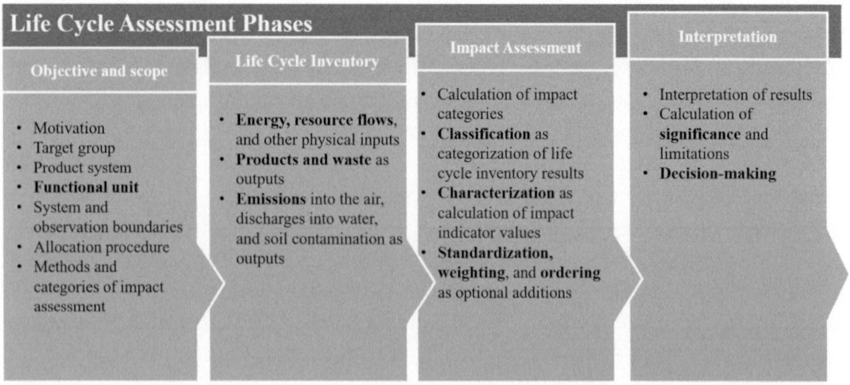

Fig. 5.4 Phases of the life cycle assessment in accordance with ISO 14040 [62]

system, and all methods and categories of impact assessment used are defined. In the following Life Cycle Inventory, the examined process chains are visualised, and the energy and material flows within the system boundaries are integrated. Thus, for the individual process steps, inputs as well as (partial) products, waste, and emissions into air, water, and soil are balanced as outputs. This type of data collection allows concrete indicators to be calculated, normalised, weighted, and ordered in the phase of the Impact Assessment. Within the impact assessment, a basic distinction can be made between problem or midpoint indicators and damage or endpoint indicators. In the literature, for example, the midpoint indicator "climate change", given in CO_2 equivalents, is widely used.

To further illustrate the above, Fig. 5.5 shows the structure of the impact paths between the mid and endpoint levels of consideration.

The resulting information is brought together in a final interpretation and examined for the significance of the solution. The user-friendly execution of an LCA can be supported by software, such as Umberto LCA+, GaBi, openLCA, or SimaPro.

The software solutions Umberto LCA+ and GaBi (by Sphera) offer powerful tools for conducting LCAs with extensive Databases and Integration Opportunities. Umberto LCA+ is characterised by a user-friendly interface and strong visualisation of lifecycle models based on material flows. GaBi, on the other hand, offers its own extensive database and a flexible modelling structure that supports lifecycle assessments considering various scenarios. SimaPro (by PRé Sustainability) uses a variety of **impact assessment methods** (e.g., ReCiPe, CML, Eco-indicator 99) to quantify and evaluate environmental impacts, such as climate change, acidification, resource consumption, or ecotoxicity. openLCA is an open-source software that supports a wide range of databases, such as Ecoinvent and ELCD, and is applied primarily in research and small companies due to its availability and cost efficiency. The methodological differences between these tools lie in the flexibility and method of modelling, the possibilities of database integration, and the resulting user-friendliness.

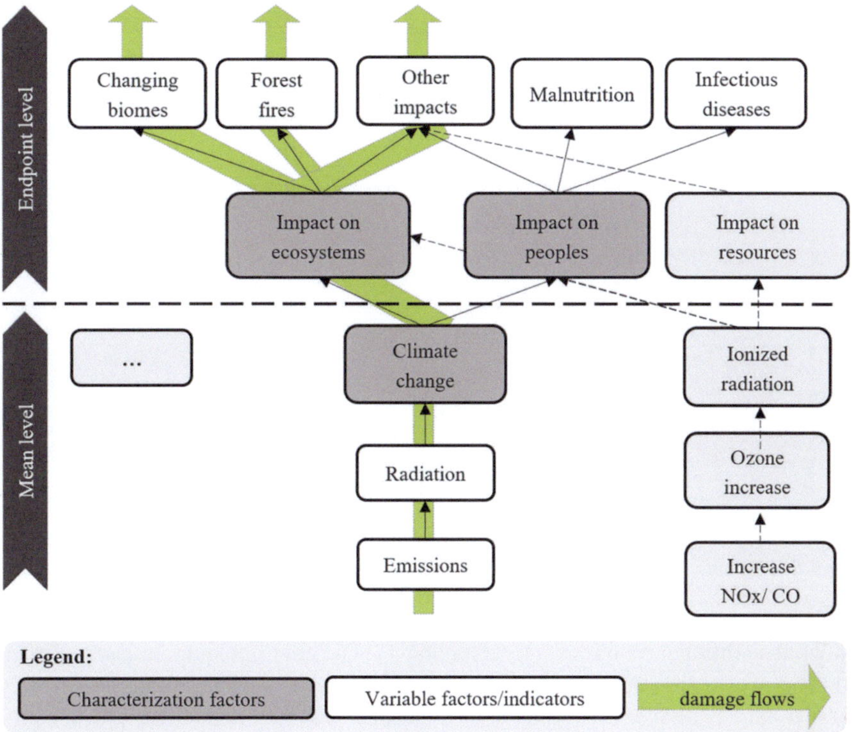

Fig. 5.5 Impact assessment—impact paths and levels of consideration; own representation based on Huijbregts et al. [63]

The following presents an example of analysing the environmental impacts of metallic additive manufacturing using PBF-LB/M (German: "Powder Bed Fusion by Laser Beam for Metals"). Tools for graphically and logically describing the cause-and-effect relationships within the examined process chains are network structures, such as material flow networks [64]. Based on Petri nets, material flow networks allow the construction of networks by linking places and transitions. A transition (T) represents a material transformation process, a place (S) represents a storage, and a flow relation (F) represents the link through materials and energies. Within the places, a distinction can also be made between inputs and outputs and storage locations. Storage locations transfer material flows between the transitions, whereas inputs and outputs are at the beginning or end of the system under consideration. Using the triple of the quantities of places, transitions, and flow relations, a process chain (N) can be described in a holistic form [64].

The conducted study discusses an ecological assessment of the MAR/R process ("Metal Additive Repair/Refurbishment"). Such methods not only support the repair of metal components but also their adaptation to new requirements or technologies. PBF-LB/M is particularly suitable for complex structures and can

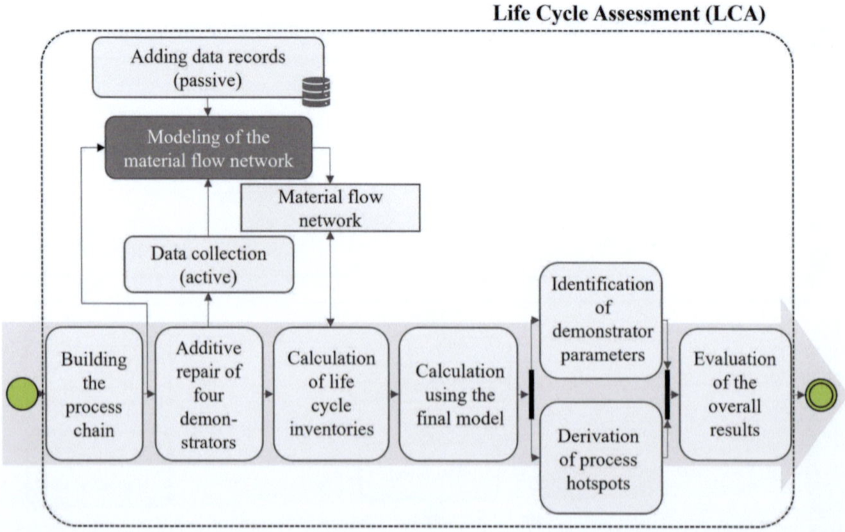

Fig. 5.6 Methodical approach to conducting an LCA according to [65] for the MAR/R process

integrate additional functions into components. The environmental impacts of such additive repair processes need to be thoroughly investigated, as there is currently limited data available. The study shows that raw material extraction, powder production, and post-processing are the main sources of environmental impact, with powder volume and processing time being particularly decisive. Detailed results of the LCA are uploaded to the repository (Zenodo) and described in Wurst et al. [65].

For the process under investigation, a methodical approach according to [66] was first established for the actual implementation of the LCA. The result is graphically illustrated in the following Fig. 5.6.

Eco-efficiency

Future research could focus on the reuse of materials, the impacts of different materials and support structures, and logistics processes to better understand the environmental impacts of MAR/R compared to new part production. Ultimately, the study suggests that MAR/R with PBF-LB/M can offer significant environmental benefits, provided the processes are designed efficiently to minimise material consumption and post-processing.

Another approach to enable an analysis and subsequent synthesis of the ecological and economic sustainability of products and processes is the investigation of eco-efficiency. As a relationship between economic value creation and ecological damage creation, eco-efficiency provides a basis for organisations to evaluate the tension between the economic and ecological dimensions. According to the original definition of eco-efficiency by [67], "eco […] efficiency is the measure of environmental impacts per unit of output". In the organisational or industrial context, the extension of the performance concept to the measure of resulting value creation provides significant support in making far-reaching decisions in the early phases of product development [68]. Despite this concrete description of the benefits, there is no universally applicable description of application due to the versatile and often organisation- and product-specific evaluation methods [69]. Rather, the approach of eco-efficiency is to be understood as a guiding principle for aligning and optimising the entire organisation and all processes contained therein to achieve the overarching goal of a more sustainable development.

6 Creativity Techniques

The following chapter deals with essential activities in developing sustainable products, namely analysing tasks and problems, building knowledge, and the necessary synthesis or compromise finding to arrive at solutions or innovations. Looking at the entire development process, numerous obstacles or barriers are encountered that need to be overcome. These can be of a technical nature but can also be subject to other necessities or obligations of internal or external company nature, such as costs, time, laws, customer relationships, sustainability criteria, eco-efficiency requirements, etc. To overcome these barriers and progress continuously and efficiently, both analysis and synthesis steps are necessary. In particular, synthesis steps explicitly characterise the engineering work to find solutions. To successfully carry out analysis and synthesis, it is worthwhile at this point to consider the basic problem-solving behaviour of humans. Solution-finding processes are initially carried out individually; however, exchange and motivation within the team can be innovation-promoting and inspiring and are recommended in many cases to overcome blockages and frustration. Based on the Design Thinking process, the steps of the described problem-solving are listed chronologically in Fig. 6.1.

The listed steps follow an iterative approach and consist of six sub-aspects in total. First, in the step *Understand*, the problem space is defined. In the step *Observe*, impressions are collected, and empathy for the users is developed. In the step *Define Perspective*, the gained insights are bundled, and the perspective is clearly outlined. This is followed by the step *Find Ideas*, in which a multitude of solutions are developed and then focused. In the step *Develop Prototypes*, theoretical models are created that can be tested, for example, with computer support. Finally, the *Testing* takes place to further detail the solutions or to specify further needs [71]. It should be explicitly pointed out at this point that the concept of

R. Lachmayer et al., *Methodology for the Development of Sustainable Products*, https://doi.org/10.1007/978-3-662-72442-2_6

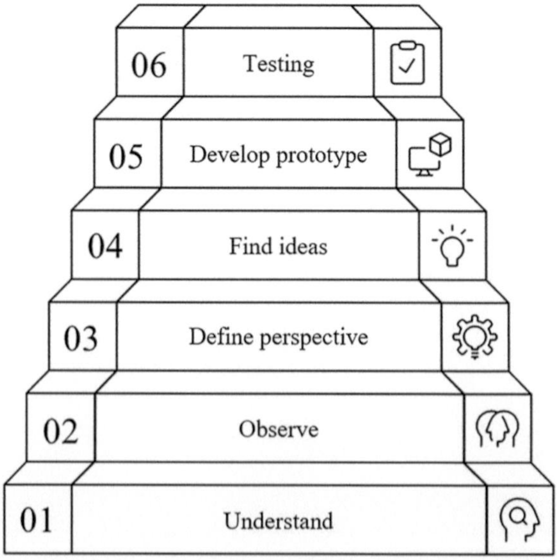

Fig. 6.1 Steps of solution finding by humans according to HPI Academy [70]

Design Thinking is not a development process, but a structuring description of the approach to solution finding.

In the context of problem setting and solution finding, the chapter should start in Sect. 6.1 with a consideration of research methods, i.e., those tools that support the finding of information and possibly also solutions from repositories. These include, for example, patent, material or process research, the search for suitable standards and guidelines, but also the classic literature search. Section 6.2 will discuss intuitive methods that promote free creativity, such as brainstorming, the Method 635, the 6 Hats Method and bionics, using examples. The chapter will conclude with Sect. 6.3, in which the discursive methods important for methodical development are presented. These have an analytical and strategic focus and are characterised by the fact that the problem is analysed in detail and solutions are sought step by step. In this section, the method of morphology and the method of construction catalogues are presented as examples. Special attention is paid to the method of construction catalogues, as several catalogues were developed in the underlying research projects that focus on the development of sustainable products. These catalogues will be explained in more detail below.

6.1 Literature Search Methods

Almost all available information is now in digital form, either in closed networks (internal to the company) or in open networks (external to the company) such as the Internet. However, a distinction must be made between free and fee-based services. Every search, therefore, begins with the question of which are the suitable platforms/services on which to search. To clarify this question, experience or communication is necessary, and language barriers may also need to be overcome. New platforms and services are continuously being created, some of which also compete with each other and are not necessarily all serious. On the other hand, existing platforms are constantly evolving, both technologically and in terms of their content, keywords, prompts, or metadata. After the question of where to search, the challenge is to clarify which technologies the respective platform offers. The next step then depends on the vocabulary—here it is necessary to define the technically correct and specific keywords and to narrow down the design space via these. During the research, the snowball system can be used, which means that further sources are inferred from the cited literature of the found sources. This also creates something like a map of the contributing institutions to the topic. Subsequently, the found information must be related to the question and evaluated in a targeted manner as well as critically assessed for its quality to contribute to the solution of the problem. This is not new as such, but it presents itself anew in the course of digitalisation. On the one hand, information and data are characterised by different maturity levels of their processing, which can be easily discussed using the FAIR Data Principles of the German Research Foundation. In this context, data should be Findable, Accessible, Interoperable and Reusable. On the other hand, the search is supported by digital tools, such as full-text searches and image recognition. Here, data are not only texts and images as in the past, but can also be digitally processed artefacts, programmes, documents, films or even 3D animations.

Another focus here is on research methods and the use of Artificial Intelligence (AI). Even though AI methods are not yet fully capable of completely mapping problem-solving processes, they can at least partially be used in a supportive role, coached by humans, to effectively assist in phases such as information gathering or visualisation of ideas. There are many methods and tools available for accumulating knowledge and finding solution ideas. It is important to note that the ongoing development of AI is bringing about significant changes in scientific literature search. By using AI-based tools, research methods can be expanded and optimised with entirely new aspects. Various AI tools offer users new approaches to identifying, analysing and networking relevant literature. Despite the many advantages, there are also limitations and challenges in using AI in literature search. In order to achieve as accurate and targeted results as possible, data clarity must be available and of high quality. In addition, aspects of data protection, ethics, and copyright must be considered to avoid negative consequences and not to question the integrity of scientific research. Based on current developments, however, it is clear that the use of AI-based tools in literature search is a forward-looking development that can promote scientific progress and help users to design results and work steps efficiently [72]. Figure 6.2 shows an exemplary approach to an AI-supported literature search.

A search, as shown in Fig. 6.2, inevitably quickly leads to many different sources. These need to be analysed, evaluated and structured in the aftermath of the AI-supported literature search.

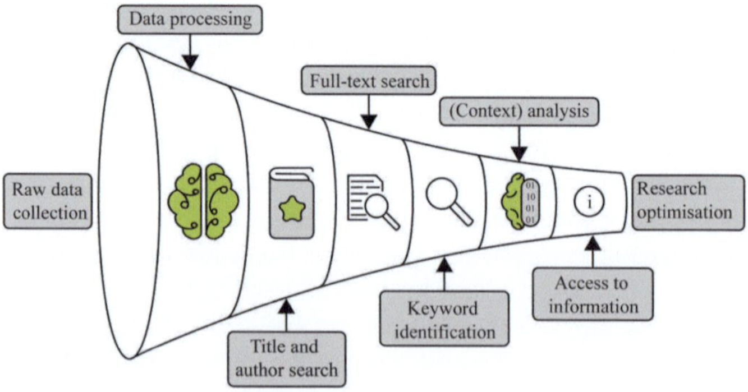

Fig. 6.2 AI-supported literature search

For many challenges, it also gets significantly more concrete. In addition to the general approach described above for systematic literature search, explicit possibilities or tools for AI-supported literature search are listed in the following Table 6.1, according to da Silva Cardoso et al. [72].

Table 6.1 Exemplary selection of usable AI tools for literature search and categorisation according to da Silva Cardoso et al. [72]

Category	Description	Tools
Assistance systems	Support, especially at the starting point of a research project Possible assistance in topic finding or formulation of research questions	ChatGPT Google Gemini HuggingChat Microsoft Copilot
Analysis tools	Support the analysis of uploaded files like PDFs Enabling a discussion about the content of the documents (e.g., creation of summaries, highlighting of key statements, filtering)	ChatPDF Elicit Explainpaper Humata
Search tools	Support, especially in the search for literature Also, possible assistance in analysis of search results (e.g., creation of summaries, highlighting of key statements, filtering)	Consensus Keenious Perplexicity AI Semantic scholar
Networking tools	Based on a starting literature, connections to networked literature are sought Thematic classification of found and networked literature	Connected papers Litmaps Open knowledge maps Research rabbit

6.2 Intuitive Methods

Intuitive, often team-based, methods generate many ideas in a short time (about 30–100 individual ideas in 60 min), particularly promoting thought associations in the search for new ideas. The intuitive methods are especially characterised by the fact that they do not follow a specific order in solution finding, but the actual solutions are generated unsystematically. An important prerequisite for the successful application of these methods is a suitable composition of the working group, particularly concerning low hierarchy levels, a broad knowledge spectrum, and the willingness for open collaboration within the team. In addition to ensuring these prerequisites, the size of the team must be adapted to the respective method. The following are examples of various intuitively emphasised methods.

Checklists

A checklist is a proven method, for example, for clarifying the task at hand. In addition to functional and administrative questions, the following questions could be part of a checklist for requirement determination under specific sustainability aspects:

- Are there sustainability differentiation features that must be met?
- What legal conditions need to be considered in the future?
- What target values for the Life Cycle Assessment need to be met?
- Which materials should not be used or should be preferred?
- What requirements for the supply chain need to be met?
- What recycling/refurbishment strategies should be considered?
- What other R-strategies should be considered?
- What form of energy should be used; what efficiency should be achieved?
- Can the development and production avoid the use of fossil fuels?
- What data flows and labelling are necessary?
- What safety and ergonomics aspects need to be met?
- What is the sensible target reliability, lifespan of the product?
- Are there manufacturing processes that should be preferred or not used?
- How can the product be developed to use fewer resources? Can the same result be achieved with less material use?
- How does the product affect aspects of social sustainability?
- How can the CO_2 footprint of the product be reduced?

Brainstorming and Brainwriting

The term "brainstorming" can be translated as a flash of thought and is used to find unbiased solutions. This takes place within a group, where the individual group members inspire each other due to the group dynamics. As a result, there is a chance that some unconventional solutions will be produced. By preventing any form of criticism during idea generation, the overcoming of thought blocks and patterns of the individual group members is enhanced. Brainstorming is a fundamentally simple method that requires discipline from all participants. Typically, a group consists of about four to ten people; in a given time, ideas on a previously clearly defined topic are collected. All participants agree to handle their ideas openly (without claim to intellectual property). A discussion-leading person introduces the topic, moderates, and collects the ideas visibly on a board or on a flipchart. This person is also responsible for informing all participants about the further procedure with the collected ideas afterwards. The discussion can either be free or based on notes, on which all participants first sketch ideas to then present them to the others (brainwriting), or two structured rounds of idea finding and exchange can follow one after the other. There must be no hierarchies in the discussion, so that immediate evaluations and criticism of ideas are avoided, and the time limit of 45 to 90 min is adhered to. Further discussions, evaluations, and analyses should only take place afterwards and possibly in smaller groups.

The following list contains classic steps for conducting a brainstorming or a brainwriting:

- Define goal/problem
- Determine participants
- Preparation
- Establish rules
- Idea Generation
- Idea Structuring
- Review
- Assessment
- Documentation
- Follow-up

As described at the beginning of the chapter, some of the creativity techniques discussed here will be concretised and exemplified using selected examples that have emerged as part of a student design project. The design project aimed to redesign and further develop conventional household items to make them more sustainable, starting from their original state. The following figure illustrates the previously described brainstorming creativity technique using the example of a kettle (Fig. 6.3).

Method 635

Many of the things already discussed above about the process and openness of the discussion apply to Method 635. However, the format is characterised by a somewhat stronger structuring. The basic idea is that six selected participants each develop three ideas on the question at hand on slips of paper and pass these on to the person sitting to their right. This person supplements the initially formulated and received ideas with three of their own suggestions and then passes them on to the right again. This is done five times in total, so that all participants have supplemented each slip of paper once. Finally, the ideas are presented to the group.

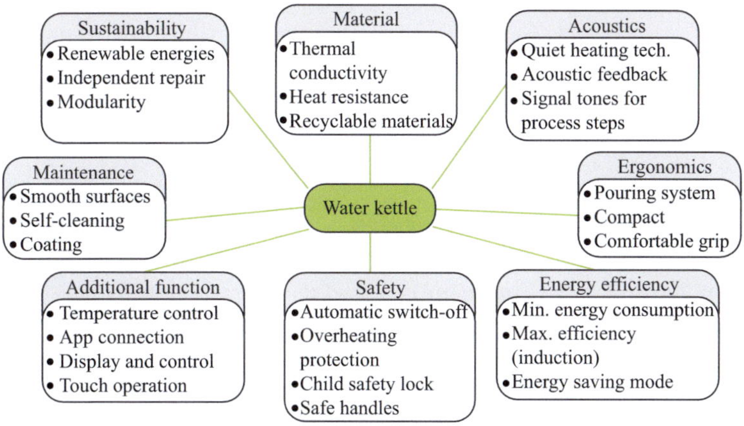

Fig. 6.3 Results of a typical brainstorming session in the development of an innovative kettle system

Review and selection of ideas, as well as definition of further proceedings, take place afterwards.

6 Hats Method

Another, somewhat different approach to an open, exploratory discussion about ideas and solution proposals is provided by the 6 Hats Method. This was introduced by Edward de Bono in 1986. The method is based on the assumption that the human brain thinks in different ways, which can be consciously targeted. By deliberately using this assumption, different ways of thinking can be induced, thus bringing different contributions into a discussion. In this process, all discussion participants are assigned a specific role via a "thinking hat". The following Table 6.2 summarises and describes how the respective roles can be defined.

Table 6.2 Roles and attributes in the 6 Hats Method

Hat	Perspective	Role description
White	Analytical	Objective, neutral, data-oriented, argumentation based on numbers and facts
Red	Emotional	Subjective, personal, feelings-oriented, argumentation based on euphoria and fear
Black	Critical	Pessimistic, objective, negative, argumentation based on risks
Yellow	Optimistic	Objective, sees advantages and opportunities, positive argumentation
Green	Creative	Provocative, innovative, value-free, argumentation rich in ideas
Blue	Overview	Structured, realistic, ordered, solution-oriented, and moderating argumentation

To illustrate the 6 Hats Method, a fictitious discussion about a wooden nutcracker is sketched out below. The participants behave according to the instructions in Table 6.2.

- **Person 1—White Hat (Analytical)**: "*The nutcracker is made of high-quality wood, making it more environmentally friendly than plastic models. It is robust and should have a long lifespan. According to the manufacturer, the wood is sustainable and from certified sources. The nutcracker weighs about 300 g and is available in various designs.*"
- **Person 2—Red Hat (Emotional)**: "*I find the nutcracker truly charming, and it evokes nostalgic feelings in me. Wood has a natural and warm quality that I particularly like. It takes me straight back to my childhood when we always used such nutcrackers at Christmas.*"
- **Person 3—Black Hat (Critical)**: "*There are definitely some concerns. A wooden nutcracker could quickly break if not handled properly. Wood can develop cracks over time, especially when exposed to moisture or constant temperature fluctuations. Moreover, the price could be significantly higher compared to a nutcracker made of plastic.*"
- **Person 4—Yellow Hat (Optimistic)**: "*The wooden nutcracker definitely has many advantages. It is sustainable and environmentally friendly, which is becoming increasingly important. Wood not only looks beautiful but can also be highly durable if properly maintained. Moreover, it is a handcrafted product, which gives many people the feeling of owning a unique and high-quality product.*"
- **Person 5—Green Hat (Creative)**: "*Perhaps we could further improve the nutcracker. What if we make it multifunctional? For example, with an additional wrench for nuts with harder shells. Or we could offer a customisable wooden surface on which various designs or engravings can be applied, making it a personal gift.*"
- **Person 6—Blue Hat (Overview)**: "*We have now illuminated various perspectives: The facts, the positive and negative aspects, as well as creative ideas for improving the nutcracker. A wooden nutcracker can be more environmentally friendly and aesthetically pleasing, but it also brings certain challenges, such as the maintenance of the material and the higher costs. We should consider whether we can find a solution to minimise these weaknesses, perhaps through better protective coatings or innovative design approaches.*"

Bionics

The principle of bionics will be examined in more detail in the following. This is a special form of synectics. This creativity method addresses and promotes the overcoming of barriers through targeted analogy formation after an intensive examination of the problem. Alienation techniques are used to break participants out of entrenched thought patterns. The alienation and analogy formation must always be followed by a retransformation and synthesis. Examples for better understanding are:

- Umbrella/Sunshade
- Heat/Cold insulation
- Radiator/Air conditioner
- Hairdryer/Dryer hood

Bionics (a word formation from biology and technology), in this context, deals with the transfer of phenomena from nature to technology. A well-known example is Leonardo da Vinci's idea of transferring the flapping flight of birds to a flying machine. A day-to-day example is the Velcro fastener inspired by burrs. Bionics is based on the assumption that living nature has developed optimised structures and processes through evolutionary processes, from which humans can learn, and which can provide more sustainable solutions in the context of sustainability.

As an interdisciplinary field, bionics attracts researchers from the natural sciences, engineering, architecture, philosophy, and design. Bionics is about the systematic recognition of solutions from living nature—it thus distinguishes itself from purposeless nature inspiration. The goal is always a technical object or process that is separate from nature. This distinguishes bionics from disciplines such as biochemistry or biophysics, which utilise and expand biological processes. In Fig. 6.4, a procedural model for product development according to the principle of bionics is shown.

By applying this approach, principles of bionics can be applied to technical products. For example, the surfaces of components can be structured with bionic shapes and geometries. Following the structure of human bones, for example, hip

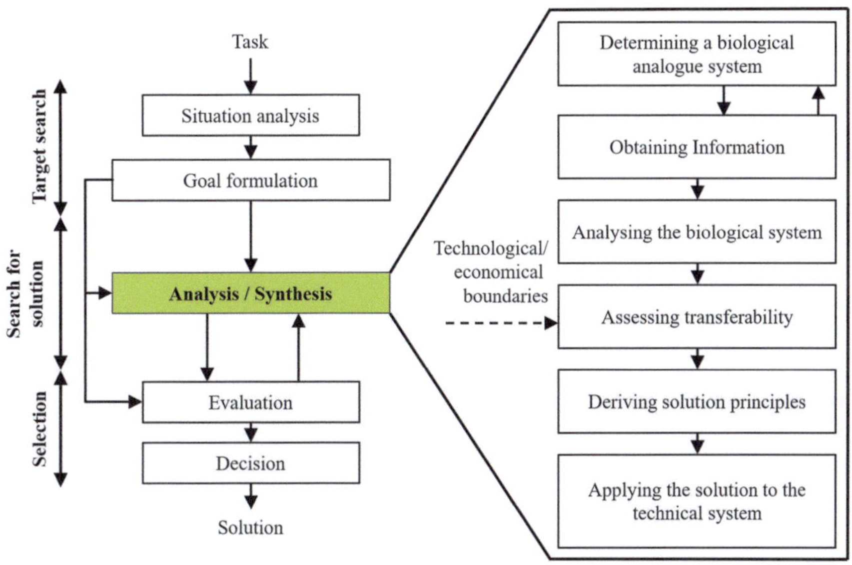

Fig. 6.4 Procedural model for product development according to the principle of bionics, after VDI 6220 [73]

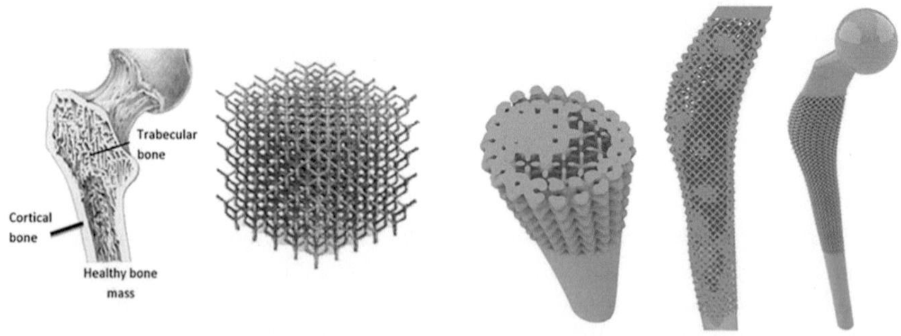

Fig. 6.5 Additively manufactured hip endoprosthesis. Left: Reconstruction of bionic surface from bony structure according to Popov et al. [74]; right: Additively manufactured prosthesis with structured surface after bionic reconstruction

implants can be designed with such bony, bionic structures and manufactured additively. As a result, the ingrowth behaviour and stability can be optimised, while less material is needed for production than for a conventional implant. The discussed example is shown in the following Fig. 6.5.

Another example where an optimised component shape and topology could be determined and implemented using bionic principles is described in detail in Sect. 10.1. In this example, a bicycle crank was designed with internal structures and a force flow-optimised embodiment, resulting in various advantages over a conventional component shape. The method of topology optimisation used is particularly suitable in the context of bionic embodiment design. Details on this method and further explanations, both theoretical and related to the example of the crank, can be found in Sect. 10.1.

Beyond the examples presented here, there are numerous other bionic solution principles, which can be found in freely accessible online databases. One example is the platform *Ask Nature.* [75].

6.3 Discursive Problem-Solving Methods

After focusing on intuitive creativity methods in Sect. 6.2, methods that generate many intuitive and quick ideas in a short time, the focus in the following section will be on so-called discursive methods. Discursive methods emphasise a slower, but above all more systematic and orderly problem-solving approach, which is primarily based on conscious and rational analysis as well as clear structuring and argumentation. Creativity methods of this discursive type generate around 10–60 ideas in 30 min. The process of finding solutions is characterised by logically sequential steps (discursive = logically progressing from concept to concept). A problem is fully described in the context of a discursive problem-solving approach by analytically breaking it down into small units. The discursive problem-solving methods of morphological analysis and construction catalogues will be presented for closer examination.

Morphological Analysis

Morphological analysis is a discursive method to fully capture complex problem areas and to consider various, potentially possible solutions without prejudice. It is carried out in a group of up to seven people, thereby increasing the knowledge and idea potential and enhancing creativity in finding solutions. The process is guided by a moderator and lasts about half to two hours. Along with the analysis of the problem, a generalisation of the question is useful. In this context, the problem field is expanded to find original solutions. The analysis uses the morphological box, which is presented as a vivid image of a multi-dimensional matrix according to the astrophysicist Fritz Zwicky [76]. This usually graphically prepared matrix forms the core of the morphological analysis.

The respective determined features (also called attributes, factors, parameters and dimensions) are defined for a specific question and listed in a column. The features should be considered independently of each other and evaluated with regard to the task at hand, in terms of their actual feasibility (operationalisability).

Subsequently, the rows of the respective features are supplemented by all possible manifestations. This results in a matrix that produces a theoretically possible solution through every possible combination of the manifestations of all features. In the specific case, combinations are set up by choosing a manifestation of a feature from each row. This selection process is repeated several times for several solution combinations. The resulting combinations of manifestations are transformed into ideas and further developed.

In the context of a specific usage of morphological analysis, it is particularly interesting in the environment of developments to break down the considered main functions into sub-functions and to develop potential solutions for each of these sub-functions. The sub-functions are written one below the other, while the associated partial solutions are assigned row by row. It can happen that there are different numbers of possible solutions. The following shows a possible structure of a morphological box in generic form according to Zwicky (Fig. 6.6).

By combining the different partial solutions, many different solution variants can be created in a short time. Rules such as clarity, simplicity, safety, sustainability, not unnecessarily converting energy flows, not taking forces for a walk, etc., can contribute to identifying one or more sensible and possible solutions.

Following the already outlined procedure, an example should also be inserted at this point. The following morphological box comes again from the previously introduced student design project and deals with the underlying theme of the sustainable redesign of a kettle. Again, the example does not provide a complete picture of the method, but it does show a possible and exemplary selection of possible results.

Following the partial solutions from Fig. 6.7, various solution possibilities can be formed. As an example of a sustainable and modern variant, an innovative kettle could be designed that is intended as a stand-alone device, is made of stainless steel

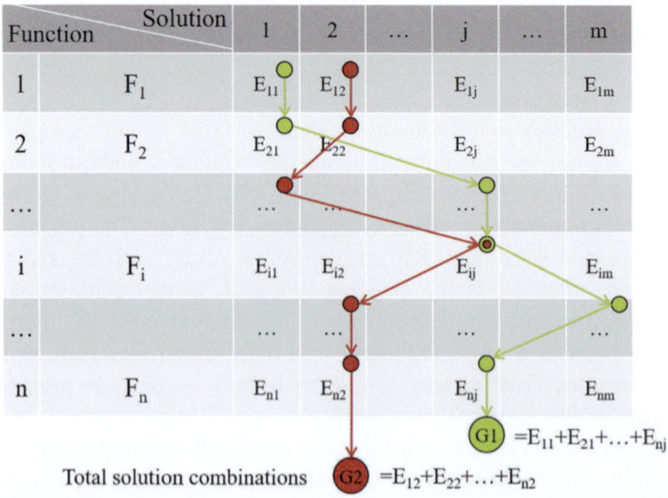

Fig. 6.6 General structure of a Morphological Box; own representation based on Zwicky [76]

for improved durability and is particularly energy-efficient as the heating process is carried out inductively. A volume of 1.5 L would be the standard here, while precise temperature control is advantageous for efficient water heating processes, so that unnecessary energy is not consumed. The electricity in this example would be sourced from renewable energy (solar power), and the kettle would be equipped with an automatic shut-off for enhanced security. An additional feature would be an integrated tea water function, so that water can be heated precisely for various scenarios, increasing efficiency. Besides this example, as can be inferred from the previous discussions, there are various other solutions, many of which are also on the market.

Functions \ Solutions		1	2	3	4
1	Design	Freestanding device	Portable	Built-in (e.g. stove)	Combined with another appliance
2	Material	Stainless steel	Plastic	Glass	Ceramic
3	Heating technology	Rapid boil plate	Heating coil at bottom	Induction	Fire
4	Capacity	0.5 L	1 L	1.5 L	10 L
5	Temperature control	Yes	No	Preset levels	Precise temperature control
6	Power source	Electric (plug-in)	Battery	USB	Solar-powered
7	Safety features	Automatic switch-off	Boil-dry protection	Overheating protection	Child safety lock
8	Additional function	Tea water function	Keep-warm function	Bluetooth control	App control

Fig. 6.7 Morphological box for a kettle

Design Catalogues

The knowledge of individual processing persons, for example, about physical effects or possible manufacturing processes, contributes significantly to the successful processing of a development task. In many cases, however, designers only have limited, sometimes highly specialised knowledge in specific areas. In order to be able to handle design tasks holistically, further knowledge must therefore be acquired and made available. Collections of knowledge, in which information or even ready-made partial solutions are structured, represent a predestined source in this context.

In the following, such types of knowledge collections are presented and examined in more detail using design catalogues. In the further course, design catalogues are referred to as information storage that is tailored to methodical designing in terms of their content, their access possibilities and their structure. Design catalogues primarily support the knowledge of designers by providing solutions for recurring subtasks and stimulating the search for further and alternative solution possibilities. Their special features are extensive completeness, a clear systematic structure and the existence of access features. The method of design catalogues or "designing with design catalogues" presented here essentially goes back to K. Roth [35]. To facilitate the use of design catalogues, they are uniformly structured. They basically consist of a structure part, a main part and an access part, and optionally an appendix. Figure 6.8 shows the schematic structure.

The **structure part** of a design catalogue determines the systematic order of the content and the completeness of the catalogue. It should be complete when setting up a catalogue and also form the framework for theoretically possible, but currently unknown, content. The structure part contains exclusively the structure features. These structural features are preferably countable or logical/structural features that divide the content of the main part clearly, largely completely and without contradiction.

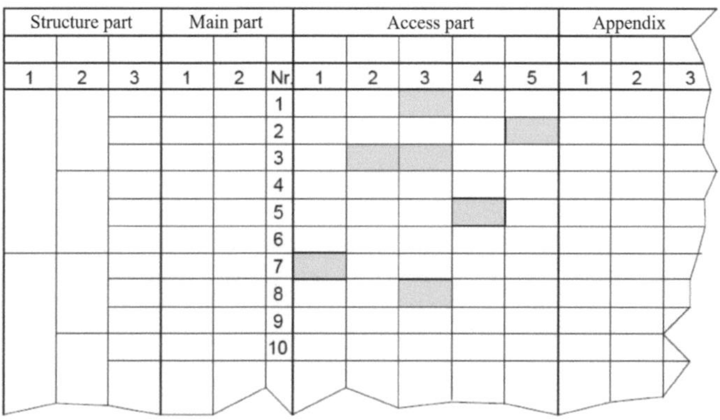

Fig. 6.8 Schematic structure of a design catalogue

The **main part** of a design catalogue contains the explicit content of the catalogue, i.e., the objects, the concrete solutions or the operations. The content is presented as clearly as possible in the form of terms, sentences, symbols, formulas and sketches or drawings. The structure features may result in fields for which no known technical solution, objects or operations are currently available. These fields are also called white fields and serve as a creative starting point for the search for new solutions.

In the **access part** of a design catalogue, the assignment of access features to the catalogue contents is organised. The access part can be adapted to the needs of different use cases and can be extended as required. Basically, the access features describe the properties of the solutions, objects or operations that are relevant for the selection of the contents during the design process.

In addition to these explanations, Fig. 6.9 shows an example of a design catalogue from the context of additive manufacturing and repair of products in excerpts. The catalogue itself can hardly be fully depicted here due to its size. For completeness, reference is therefore made to the research data repository of Leibniz University Hannover, in which the complete data can be digitally accessed and viewed (https://data.uni-hannover.de/de/dataset/use-case-katalog-metal-additive-repair-refurbishment).

The catalogue contains the elements generally applicable to design catalogues, as defined earlier: the structure features, the main part, the access part and the appendix. The structure part of the catalogue arranges various use cases of Metal Additive Repair/Refurbishment (MAR/R) based on the respective initial state

Structure part		Main part					Access part				
		Component						Dimensions			
Initial state of component	Type of damage or failure, outdated component area	Label	Machine/ Sector	Picture	Source		Material	Outer dimensions: $L \times W \times H$	Component area to be manufactured with thin-walled structures	Component area to be manufactured	Component area to be manufactured
1	2	1	2	3	4	#	1	2	3	4	5
Damaged	Corrosion										
	Wear & tear										
	Fracture										
	Crack										
	Deformation										
Manufactured incorrectly	AM production failure										
	Production failure by cutting / post processing										
Outdated	Internal channels and exit openings										

Fig. 6.9 Excerpts from the design catalogue "Metal Additive Repair/Refurbishment" according to Ganter [66]

of the components that led to the execution of the MAR/R process chain. For damaged components, a categorisation of damage patterns was developed in the structural part, which is based on the VDI 3822 guideline. This categorisation divides the damage patterns into corrosion damage, wear damage, breakage, crack, deformation and malfunction due to deposition. The main part of the catalogue contains the name of the component, the machine or industry in which it is used, a picture and the source. In the access section, component specifications are provided. Users can thus find similar MAR/R use cases for their damaged part based on the component features in the access section or the initial state in the structure section. In the columns following the access section, the MAR/R process chain is documented. The goal (repair only, modernisation only, or repair and modernisation), the process steps carried out, and the technologies, machines, and materials used are specified, if known. [66].

With the help of the described design catalogue, use cases for the repair and/or modernisation of components using additive manufacturing methods can be summarised. It serves as a support for users to select the appropriate additive manufacturing methods for an existing part and to determine the next steps in the process chain.

For a general understanding of design catalogues, it should also be noted that they differ from each other based on their content and are consequently distinguishable from each other. Depending on the respective application purpose, design catalogues are differentiated into solution catalogues, object catalogues, operation catalogues, and relationship catalogues. Relevant details are briefly explained in the following list.

Solution catalogues are design catalogues in which a direct assignment of functions to tasks or specific effects to effect carriers takes place. Solution catalogues thus always contain constructive solutions for a specific task. The solutions can be described at different levels of abstraction, for example, effects for force generation or geometric representation of effect carriers, but the underlying task remains unchanged.

Object catalogues are design catalogues that contain task-independent and fundamental facts, such as geometric, physical, technological, or material science information. These facts are of general interest for designing. Classic examples of object catalogues are collections of manufacturing processes, moments of inertia for important bodies, or similar part assemblies such as ball bearings, seals, or forced transmissions.

Operation catalogues are design catalogues in which procedures and rules, for example, for the generation, linking, or selection of solutions, are summarised and systematically arranged. The respective contents are always related to specific objects. Typical operation catalogues contain rules for the variation of functional structures or the calculation of tolerances.

Relationship catalogues are design catalogues in which relations between contents or functions are depicted, i.e., a relationship between two objects is represented. Relationship catalogues are seen as a special case of solution catalogues, in which a limitation to two specific objects is made.

In addition to numerous design catalogues, which can be found especially for general technical questions, for example, in Roth [35], many catalogues also contain company-specific know-how, which is not generally accessible. To distinguish design catalogues from other collections of solutions, three essential properties can be noted:

- Design catalogues allow quick access to information and are valid for a large number of users.
- Design catalogues are adapted to the necessities of the respective design process, largely complete, and free of contradictions.
- Design catalogues are clearly structured and expandable.

To ensure these properties, a structured and uniform approach to the creation of design catalogues is sensible and necessary. A generic approach to the creation of design catalogues is shown in Fig. 6.10.

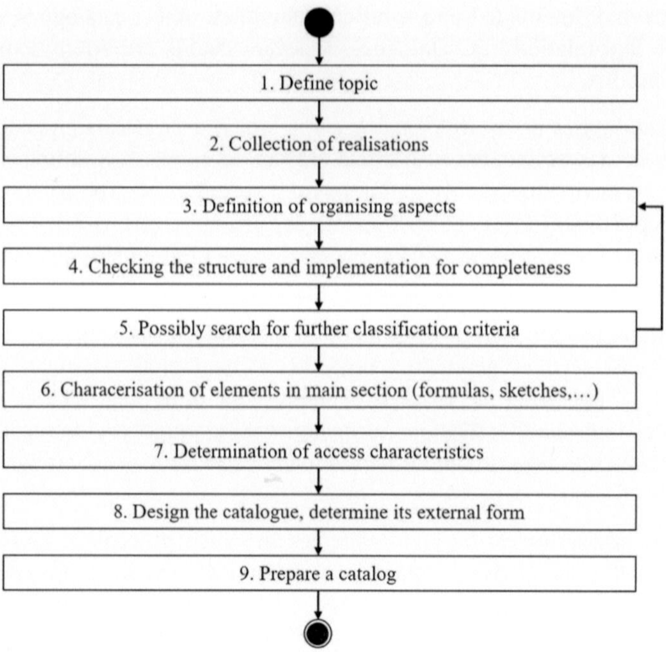

Fig. 6.10 Steps for creating a design catalogue based on Roth [35]

7 Task Clarification and Requirements Management

In the discussion, it is often assumed that developers are considered particularly efficient when they do not extensively debate basic questions and problems, but rather present concrete solution proposals promptly. This approach allows them to primarily deal with complex, detailed questions—preferably without further discussion of basics. However, it is often overlooked that in most cases a comprehensive and well-founded concept is required, as even a detailed elaboration of details cannot compensate for an insufficient or non-existent overall concept. To address problems of this kind, a clear clarification of the task is necessary.

The goal of such a task clarification is thus to define the required product properties and to describe the framework conditions for the production and use of the product. In the early stages of product development, increased care is required, as errors made in this phase can hardly be corrected later on or only with considerable effort. This also applies in particular to the environmental impact and legal conformity of a product. In later phases of the product life cycle, these can hardly or not at all be remedied as a result of potential omissions in early development phases and can therefore represent considerable risks and restrictions for the sustainability of a product.

In general, it should be noted that development and design tasks have different origins. Essentially, they can be traced back to:

- An internal development order funded by the company itself, based on market needs and product planning results;
- Legal requirements;
- The specific order of a particular solution by a customer;

Suggestions for quality improvement and thus cost reduction.

R. Lachmayer et al., *Methodology for the Development of Sustainable Products*,
https://doi.org/10.1007/978-3-662-72442-2_7

7.1 Task Clarification

Task clarification is carried out, based on the initiatives mentioned above, by the various departments working on the project. It includes the collection of information on functions, neighbouring systems and conditions, a vulnerability analysis, competitive analysis, and the analysis of the life cycle phases of the product to be developed.

The following questions can be used for information collection:

- What physical relationships can describe the problem?
- What physical relationships occur in the environment of the new system/product?
- Into which overarching systems is the product integrated?
- Can initial solution ideas already be generated, or are they even already available?
- What non-technical influences affect the new product?

Describing the task through existing physical relationships poses a significant challenge for developers in new developments, while in further developments, in the context of technical inheritance (see Chap. 4) across product generations, many similarities often exist.

The description of a task can also be based on existing interactions between a product and its environment. In addition to describing the system under consideration, its effects as well as its temporal and spatial relationships are also considered. The following environmental systems can be described:

- Passive system, on which the product has an effect
- Active system that acts on the product
- Command system through which the product is controlled
- Site of action with the prevailing climate system and spatial conditions
- Creation and transport systems
- Maintenance system
- Disposal system

A vulnerability analysis, also in the context of data-driven product development, is particularly effective when predecessor generations of the product are already on the market. In principle, vulnerabilities of a product generation can be identified throughout the entire product life cycle. As a result, it is important to consider all specialist domains in the vulnerability analysis. This can be achieved by either conducting expert discussions or, ideally, by having data available that can be evaluated. Aspects that can be analysed include:

- Quantities
- Product quality
- Revenue

- CAPEX, OPEX, contribution margin
- Profit, to secure new business plans through experience / derive Key Performance Indicators (KPIs) for "Target Costing"
- Market shares, market segments, and changes in market share, associated series structure, and variety of variants
- Inventory, stocks, turnover frequency, delivery times, efficiency of the process chain
- Production data, manufacturing parameters
- Technical parameters based on measurements of target achievement
- Complaints, warranty cases, and scrap rates
- Customer data, customer feedback, surveys, and reviews
- Delays, compliance, penalties
- Operational data, usage behaviour, wear patterns, performance issues
- Environmental impacts, sustainability, and sustainability assessment
- Life cycle analysis, material efficiency, energy consumption, ecological footprint
- Durability, reusability, repairability, and recycling
- Social responsibility, working conditions, and social impacts of the product

Competitive analysis, often also called benchmarking, refers to the systematic collection and evaluation of information about competing companies, their market activities, and their products or services. Based on such a competitive analysis (such as benchmarking) and the observation of political conditions or lobbying, a company can derive strategies for changing conditions and adapt existing strategies. With this approach, companies are then able to serve market gaps in a targeted manner. The above questions of the vulnerability analysis can also explicitly refer to the products of competitors. In addition, the following questions can be used as a starting point for further information gathering:

- Who are the main competing companies, what does their organisation look like, what is their legal form, and where are their locations?
- What products and services do these competing companies offer?
- Who are the market-leading companies and what goals, intentions, and strategies do these competitors pursue?
- What prices do competing companies charge?
- How do the products of competing companies differ from our own products? What are their strengths and weaknesses?
- What are the success recipes of competing companies? Is there a possibility to catch up with their optional lead?
- If so, what are the necessary expenditures for this?

The aim of analysing the life cycle phases of a product is to determine the nature of the relationships of the future product to people and natural as well as technical neighbouring systems during the individual life cycle phases. As a result, it should be possible to derive explicit requirements from this. All people and systems that are in a relevant relationship with the product in any life cycle phase can be considered part of the product environment. The term relevance implies that it is first necessary to limit the interactions considered relevant, which is usually not a simple process. Resulting interactions arise from material, energy, and information flows (see Chap. 4) as well as ergonomic and safety-relevant aspects.

To support this approach, there are a multitude of search schemes that facilitate systematic analysis for developers. Table 7.1 shows an excerpt from a search matrix structured according to life cycle phases and neighbouring systems with questions for determining the desired product characteristics according to Roth [77]. A complete version of the table can be found in the initial source.

The individual aspects listed in Fig. 7.1 are initially generic and must be adapted to the respective tasks at hand. For the recycling-friendly design of batteries, for example, the life cycle phases of raw material procurement and return will need to be examined much more closely; for the production of solar cells, the life cycle phase or the production of the necessary polysilicon.

7.2 Systematics of the Solution Space

The term solution space describes the idea that there is not just one optimal solution for a product. Instead, there can be several solutions. These are located on a so-called Pareto front. Depending on the requirement or boundary condition, the solutions allow different prioritizations and compromises between criteria such as cost, quality or function. This variety of possible solutions reflects the adaptability of a product to different needs and demands. Furthermore, the necessary design space is to be expanded by planning products not singularly, but directly as scaled series (for example, by size). It should also be considered that products often have to include entire modular systems of individual solutions to be competitive.

Figure 7.2 shows the relationships in the context of a solution space systematics using the example of a process chain developed by Modum Shoes GmbH for the industrial production of custom-made shoes. The idea of Modum Shoes GmbH is based on industrially produced, high-quality, and durable custom-made shoes instead of mass-produced goods, of which about 320 million pairs are disposed of annually in Germany. The customer scans their feet with their mobile phone and sends the data. Based on this, an individual digital last is automatically created for the customer using characteristic parameters.

This digital process chain does not use standard sizes based on grading and scaling. Instead, an individual pair of shoe lasts, the central manufacturing tool in shoe production, is developed for each customer based on their body measurements and used for the production of the shoes. This order-related business model thus enables industrial and resource-saving shoe production with an individual fit. While conventional mass production of shoes causes overproduction of up to 40% [78], Modum Shoes GmbH only produces what the customer actually orders. The process chain for the industrial production of custom-made shoes significantly reduces material consumption, storage effort, and CO_2 emissions. Given that the

Table 7.1 Structuring features for a requirement list according to Roth [77]

	Main characteristic	Secondary feature	Examples
1	Function	Overall function, Subfunction	Effect, (Main) Task, Goal: Solution-neutral short version of the task
2	Geometry	Dimension, space requirement, number, arrangement, …	Width x height x length, size, diameter, expansion, position in space, dimensions, area, …
3	Kinematics	Type of motion, speed, acceleration, reference system, …	Uniform, jerky, linear, rotational, magnitude values, …
4	Forces	Force, weight, force effect, stiffness, resonances, …	Size, direction, frequency, max. values, permissible values, deformation, pressure, …
5	Energy	Power, efficiency, loss, state variables, storage, …	Demand, friction, ventilation, pressure, temperature, humidity, heating, cooling, …
6	Material	Physical/chemical properties, material, material flow and transport, …	Input and output products, lubricants, …
7	Signal	Information, signal form, display type, …	Input and output signal, analogue/digital, sound, light, …
8	Security	Protection systems, operational, work and environmental safety, emissions, …	Guidelines, product liability, …
9	Ergonomics	Human–machine relationship, design, …	Operation, type of operation, clarity, form design, …
10	Quality	Measurement possibilities, testing possibilities, regulations, …	Lengths, diameters, angles, surfaces, functional dimensions, industry standards, DIN, ISO, …
11	Manufacturing	Manufacturing process, equipment, manufacturing accuracy, …	In-house production, outsourced production, manufacturable dimensions, tolerances, surfaces, …
12	Laws	Product safety requirements, environmental regulations, occupational safety guidelines, …	Product Safety Act, General Data Protection Regulation, Occupational Safety Act, Circular Economy Act, …
…	…	…	…

clothing industry accounts for approximately 4% of global greenhouse gas emissions [79], this business model makes a relevant contribution to the development of sustainable products.

In Fig. 7.3 a picture shows various components of a shoe. The core of the innovation is a knowledge-based system that fully automates the design and manufacturing preparation of the customer-specific last .

Properties and boundaries / Phases of life cycle		a	Technical-physical		Human related		Economical		Normative
		b	Technological and functional	Physical and nature related	Physiological	Psychological	Cost related	Organizational and planning	Physical and nature related
1	2	Nr.	1	2	3	4	5	6	7
Production	Product planning Development Construction	1	1.1 State of the art, Development knowledge	1.2 Known natural laws and effects, Materials	1.3 State of ergonomics, ergonomic testing facilities	1.4 Motivation, Training of development department	1.5 Costs of development	1.6 Duration of development, Consider higher business goals	1.7 Intellectual property rights for solution concepts
	Preparation Manufac-turing	2	2.1 Available sources for production and utilization						
	Assembly	3	3.1 Available sources for assembly						
Distri-bution	Trans-port	4							

Fig. 7.1 Excerpt from search matrix for clarifying the task according to Roth [77]

A problem-centred approach was chosen, which implicitly maps the solution spaces by formalising the design process using algorithms. With reference to the dimensions of the solution space, the following three aspects result.

1. Relevant foot and body measurements: A multitude of relevant body measurements are used to calculate the control points and weights of the NURBS surfaces that define the parametric geometry of the shoe last. These include both linear measurements and circumferential measurements, and possibly other body areas (such as height or weight).
2. Model selection and design preferences: Potential customers can select various shoe models in the ordering process, which differ in terms of design language (for example, toe shape), chosen materials, and the pattern used.
3. Manufacturing parameters: Depending on the chosen manufacturing site or the desired model, different requirements arise for the embodiment design of the respective individual parts. These affect a multitude of parameters along the process chain—such as variably defined seam allowances, material or machine-specific tolerances, and manufacturing-related adjustments of the geometry.

Another aspect of the solution space to be mentioned is time. This can be taken into account, for example, by planning several product variants from the outset. This creates the opportunity to promote and place product variants for task clarification, which makes it possible to strategically reintroduce a product to the market after several years.

With regard to the specification, the design space must be clearly defined at all times, as it significantly influences both the concept selection and all models and parameter studies to be created during development.

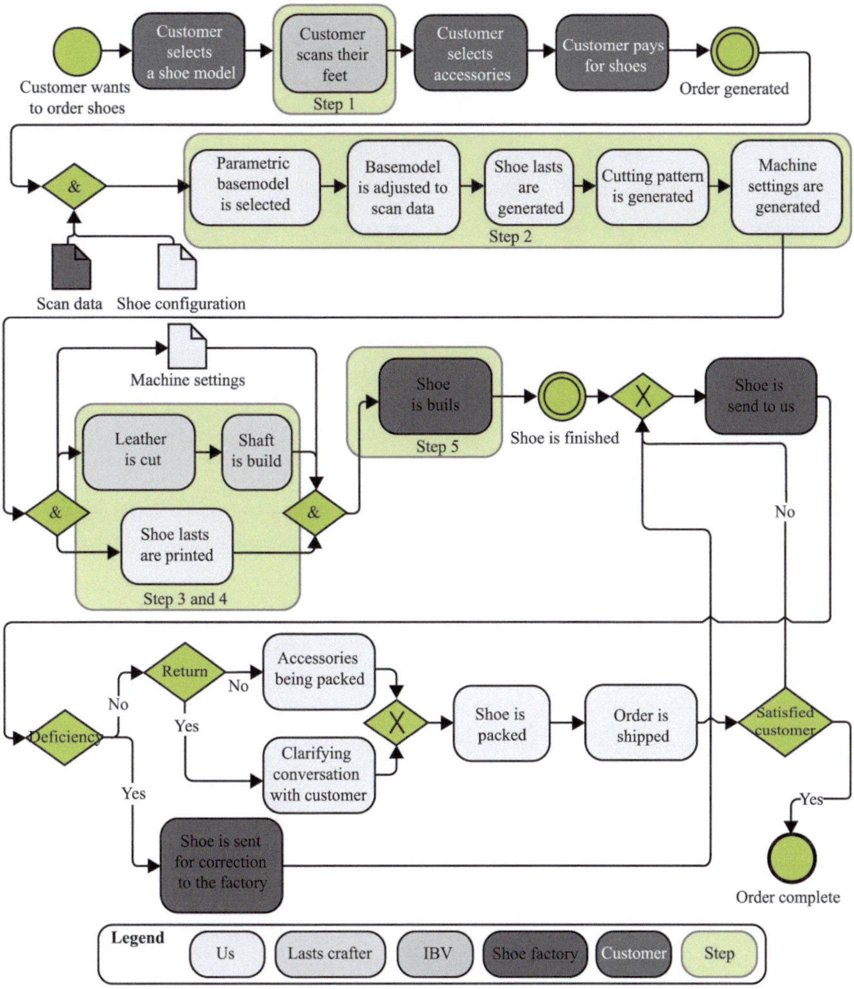

Fig. 7.2 Overall process for the industrial manufacture of bespoke shoes

7.3 Requirements Management

With the help of the presented methods for task clarification and the concept of the solution space, essential requirements for products, product families, or modular systems to be developed can be determined. However, due to the usually large number of requirements of different characteristics and relevance, these must be structured, quantified, and organised throughout the entire development process. Today, computer-aided tools such as IBM DOORS, Jama Connect, ReqView,

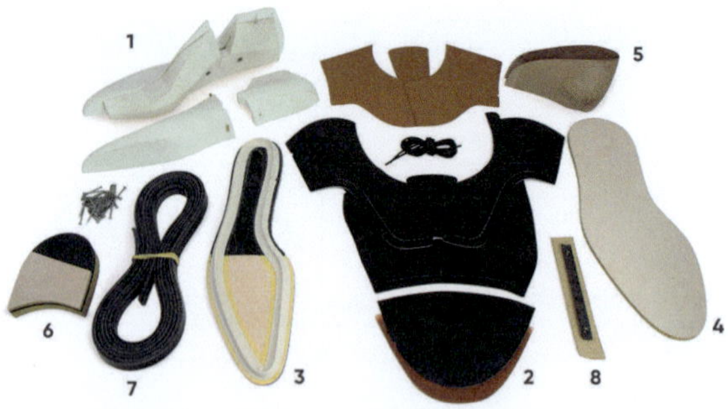

Fig. 7.3 Customer-specifically developed and manufactured individual parts: In addition to the shoe last, all other components are automatically designed by the knowledge-based system, which is also directly integrated into the manufacturing process and takes over the CNC programming and process control

Xebrio, and others are available for this purpose. However, these will not be discussed in detail in the following. Instead, the focus will be on fundamental aspects of a methodological nature for managing requirements.

The fundamental goal of requirements management is to continuously ensure the consistency and clarity of a project's objectives. These aspects must always be maintained, even as the level of detail increases. Moreover, it is of particular importance to avoid redundancies in formulating the objectives. It is both a challenge and a task for the developers to derive a compromise from the various requirements of different stakeholders and with the help of synthesis steps. Variations in the objectives within a development often lead to parts of already completed work becoming obsolete. Therefore, such changes or adjustments in the objectives should only be made for valid reasons and with thorough consideration of all potential impacts on the ongoing project.

Figure 7.4 shows the temporal dependency of the number of requirements necessary for processing the development task, the number of defined requirements, the number of documented requirements, and the intensity of consciously compiled requirements.

As can be seen from the figure, at the beginning of the development, essential properties and requirements, as well as existing relationships between these, are defined, mainly using various methods. In the case of new developments, these requirements are often unspecific and global in nature due to the initially limited knowledge. As the project progresses, the intensity of active and conscious requirement clarification usually decreases significantly, while the detailing of the product continuously increases. This is illustrated in the figure by the increasing number of requirements over time. As the product's detailing increases, so does the knowledge about the product. This leads to the formulation of new, often solution-oriented requirements or the refinement or modification of existing requirements. Often, the resulting requirements are either not documented at all or at least not fully, in the sense of requirements management.

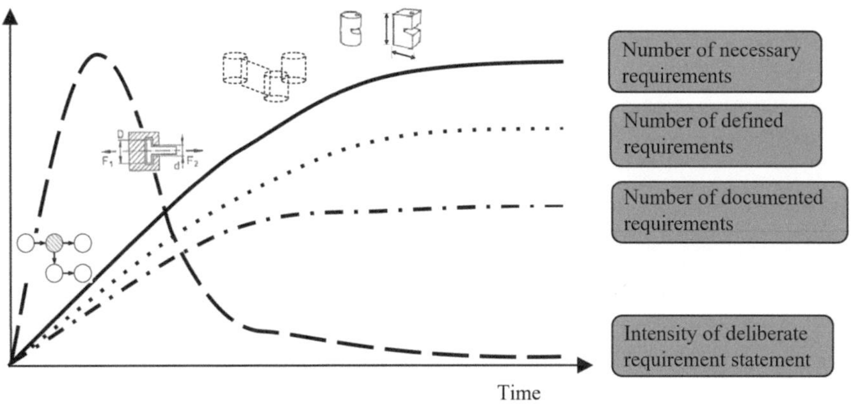

Fig. 7.4 Requirement clarification over the course of development according to Jung [80]

In the identification of requirements and the necessary requirements management, different types of requirements can be distinguished. These can be market-related, technology-related, or customer-oriented. In the following, the Kano model will be used to illustrate the influence of customer requirements on the success of a development project and how requirements can be structured and categorised in this case (from the customer's perspective). Basically, customer requirements can be divided into excitement, performance, and basic factors. Basic features are taken for granted by the customer and not explicitly expressed. Performance factors, on the other hand, are features that are explicitly stated and demanded by the customer. The fulfilment of such performance factors is seen as necessary. Excitement factors are not explicitly expected by the customer and cause excitement when present. Over time, excitement factors first become performance factors, then basic factors. The following figure graphically illustrates the discussed relationships of the Kano model.

After the general necessity of requirements management has been discussed at the beginning of this section and the exemplary execution of customer requirements using the Kano model has been shown, the question of how requirements can be structured will be addressed in the further course.

One way to systematically capture and document requirements is the so-called requirement list. This is created to systematically record all the properties, functions, boundary conditions, and restrictions identified in the previous steps that the product under consideration must fulfil. Unlike the functional specification and technical specification, which each reflect either the perspective of the commissioning company (functional specification) or that of the commissioned company (technical specification), the requirement list serves as an internal document. By creating such a requirement list, a more comprehensive and neutral capture of the formulated requirements can be enabled and realised.

A requirement list must contain at least the following elements of the basic structure for formal description:

- **Structure**: Subsystems, functional groups, assemblies, life cycle phases;
- **Number**: Identification number for unique assignment, e.g., project or part number;
- **Designation**: Clearly formulated designation, e.g., product or component name;
- **Values, data**: Quantification of requirements;
- **Type**: Binding nature of the requirements, e.g., fixed, target, or desired requirement;
- **Source, remark**: Recording of remarks, sources, and responsibilities, e.g., name of responsible employee and change statuses.

To simplify access to the information recorded in a requirement list, a content-related and process-adapted structuring is helpful. For this purpose, the structuring features shown in Table 7.1 and listed as examples can be used, which categorise and subdivide the requirement list into main features and secondary features. This type of structuring can also be found in the example shown in Fig. 7.6.

Each requirement should be numbered and formulated as clearly as possible so that the required property of the product is clearly composed of features and values, as well as associated units. Example entries in this context would be the specification of a maximum speed of 120 km/h or a minimum force of 200 Nm.

In principle, three types of requirements are distinguished. These are fixed requirements (F), target requirements (Z), and wish requirements (W), which are also identified as such in the course of a list creation. At this point, it should also be noted that there is a correlation with the previously mentioned basic, performance, and delight features of the Kano model (see Fig. 7.5).

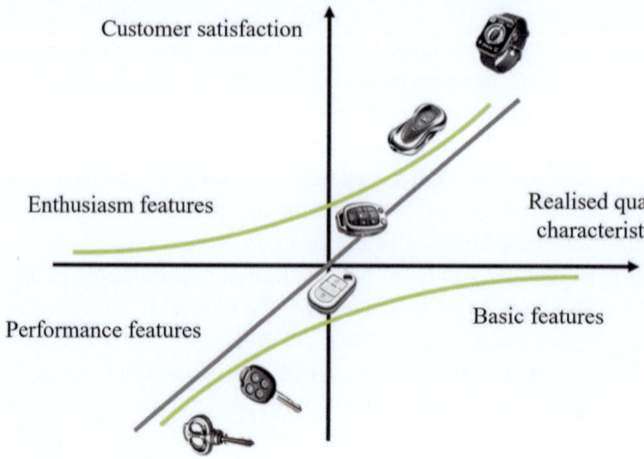

Fig. 7.5 Kano model using the example of a car key according to Heine [81]

iPeG Leibniz Universität Hannover	**Requirement specifications** **Project: Water kettle**	M = Mandatory req. T = Target requirement W = Optional req.
Created by:	Approved by:	Rev. 1
Created on:	Approved on:	Site 1/1

Nr.	Requirement	Type	Value / Range		Variable / Unit	Source / Reference
			Exact	Tolerance		
1	***Function***					
1.1	Electromagnetic compatibility (EMC)	M				Directive 2014/30/EU
1.2	No hazards for users during operation	M				ProdSG
1.3	Temperature control	O				Client
2	***Geometry***					
2.1	Overall height of the kettle	M	200-300		mm	Marked analysis
2.2	Capacity	M	1,5-2,0		liters	Marked analysis
3	***Forces***					
3.1	Weight of kettle	T	0,8	<	kg	Marked analysis
3.2	Weight of socket part	M	0,2		g	Client
3.3	Internal pressure resistance	T	1,5		bar	Client
4	***Energy***					
4.1	Rated power of kettle	M	2000-3000		W	Marked analysis
4.2	Rated power of power supply	T	230		V	Directive 2014/35/EU
4.3	Maximum operating current	M	13	<	A	Client
4.4	Low standby consumption	O	6	>		Client
4.5	Energy-saving mode	M				Client
5	***Material***					
5.1	Housing made of stainless steel	M				Client
6	***Signal***					
6.1	Acoustic signal when water is boiling	M	50-75		dB(A)	Client
6.2	Optical signal when water is boiling	O				Client
7	***Design***					
7.1	Black handle	O				Client
8	***Cost***					
8.1	Manufacturing cost	T	40	<	Euro	Client

Fig. 7.6 Excerpt from a requirement list for a kettle

- **Fixed requirements** can result from laws or safety rules as well as ethics and compliance, but can also be prescribed by the customer via the functional specification and must be strictly adhered to. Non-compliance with a fixed requirement leads to the exclusion of the solution.
- The value of a **target requirement** represents the limit that must not be exceeded or fallen short of. Through the degree of fulfilment, i.e., the distance to this limit, the quality of different solutions can be determined and compared with each other. The degree of target achievement is, especially for quantifiable characteristics, an important assessment criterion. If the defined limit is exceeded or not reached, the chosen solution variant must be adapted or completely cancelled.
- **Desired requirements** can be met, but do not have to be. However, they are crucial in everyday products, where their implementation is state-of-the-art technology. Consumers often base their decisions on implicit desires and brand image, which must be anticipated, especially in the development of consumer products, if the product is to be successful in the market.

The following excerpt of a requirement list in Fig. 7.6 comes again from the previously introduced student development project and deals with the underlying theme of developing a sustainable kettle. The list serves as an illustrative example of the theory described above.

For requirement lists of this kind, it is common practice today to create and/or manage them digitally, for example, using software-based solutions like *IBM DOORS* (Dynamic Object-Oriented Requirements System). They are a generic component of the digital tools for requirements management mentioned at the beginning.

8 Product Architecture and 9R Strategies

The following chapter discusses the concepts of product and function structure as well as product architecture. First, basic definitions of these terms are explained to provide a general understanding of the structure and functionality of products. Subsequently, various approaches are highlighted that can integrate sustainability aspects into the embodiment design and structure of products. The first approach discusses the principle of modularisation. The second approach explains considerations on the concept of R strategies and places them in the context of the chapter.

Essentially, the concept of product structure serves to represent a part in hierarchical relationships and to structure the product and its divisions. The product structure describes the physical, hierarchical composition of a product from its assemblies, sub-assemblies, and parts, as is often documented in so-called structure parts lists. The product structure significantly influences subsequent product life phases, such as procurement, production, assembly, logistics, repair, and recycling. For the successful development of products, a product structure that aligns with the goals of the entire product portfolio and the company, and especially with the specific purpose, is essential. The described relationship is exemplified in the following Fig. 8.1 using a kettle as an example.

As can be seen from the figure, the product kettle is divided into assemblies and parts at various hierarchical levels. Within the depicted product structure, individual assemblies, such as the container, and the associated sub-assemblies, such as the filler body, are summarised at a deeper level, which in turn are divided into individual parts, such as the heating plate and insulation. In the course of such a representation of the product structure, dependencies between the individual elements are also shown.

In addition to the product structure, the functions already discussed and the associated function structure are of crucial importance for a product. The function

R. Lachmayer et al., *Methodology for the Development of Sustainable Products*,
https://doi.org/10.1007/978-3-662-72442-2_8

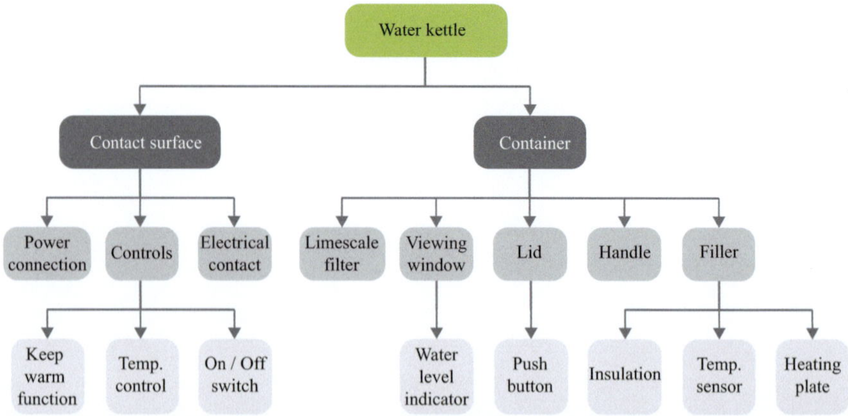

Fig. 8.1 Product structure using a kettle as an example

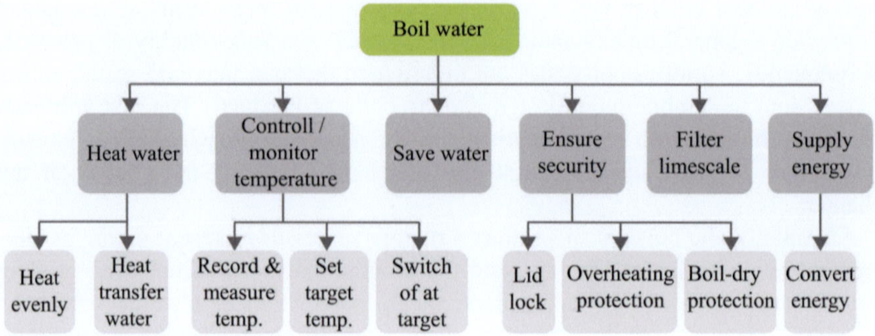

Fig. 8.2 Function structure using a kettle as an example

structure is a schematic representation of relationships between different functions and sub-functions starting from an overall function of the product. Similar to the product structure, the function structure is often represented as a hierarchical diagram with various levels of detail in the form of a tree diagram, which is also referred to as a function tree. When creating such a function tree, the overall function must first be determined. It should be noted that a product can only have one overall function. This overall function is then divided into sub-functions, and auxiliary functions are addressed. This process continues until all functions are represented. Here, too, the representation shows links and dependencies between individual functions. In Fig. 8.2, the explained relationship is exemplified, again using a kettle as an example.

As can be seen from the figure, the product kettle is divided into functions and sub-functions at various hierarchical levels. Within the depicted function structure, the overall function, boiling water, is divided into individual functions, such

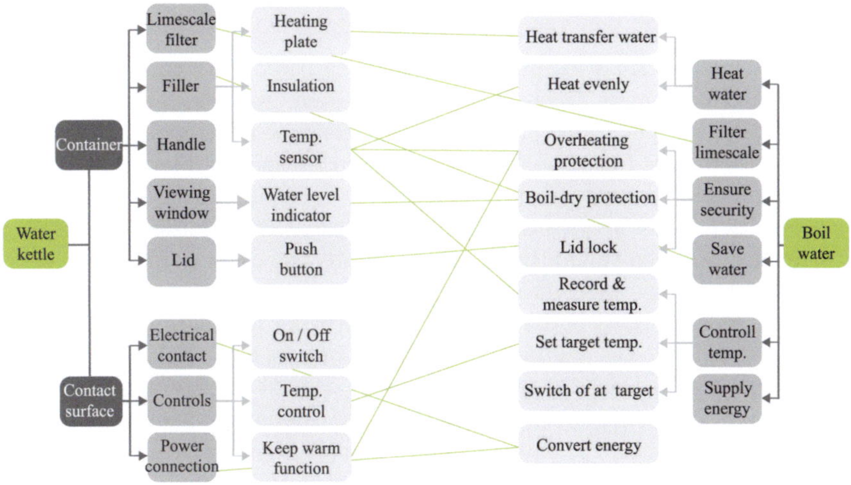

Fig. 8.3 Product architecture using the example of a kettle; own representation based on Krause et al. [82]

as temperature control, and the associated sub-functions, such as temperature measurement and adjustment.

Generally, it should be noted that product structure and function structure are usually not identical, as functions often require several components for their implementation, and conversely, a part can indeed realise several functions or be involved in their implementation. However, since function structure and product structure are inherent to the same product, there are clear correlations. If these are to be represented, the so-called product architecture is suitable for use. As previously described for the product and function structure, different levels of detail and hierarchical levels also apply in the case of product architecture. This allows correlations to be viewed either at the overall product level or at the level of individual subsystems. To create a product architecture, the product and function structures, ideally created in advance, are combined. Together, these form the product architecture. This represents the central result object of the concept phase in the development process. A product architecture can be used to define and describe in detail various construction methods, such as platforms, module strategies, common part strategies, or construction kits.

Based on the previous illustrations, a product architecture for the example of a kettle is created and shown in Fig. 8.3. While the product structure is plotted on the left side, the function structure is found on the right side of the figure. The respective elements are linked via connections and thus placed in the prevailing dependencies.

Extending the theoretical foundations presented so far, the next section will deal with the implementation of product architectures using specific designs. First, the example of modularisation or the module design will be examined in more detail.

8.1 Modularisation

Modularisation describes the principle of dividing particularly complex products into modules that are as decoupled as possible from each other. This allows for independent development and production of different modules, as well as their use in different products (platform strategy) or their combination into different products (construction-kit strategy). A common parts or module strategy can also mean that modules or parts are used that exceed the required specifications. This is done to increase production volume by using high-quality components, thereby enabling a higher unit quantity. Here, the cost optimum between scale effects and additional effort in the components is usually sought. Oversizing thus describes the fulfillment of the requirements of different application variants by a single component or product. From a sustainability perspective, other optimal solutions can be considered, as over-engineering typically leads to increased material and manufacturing effort. The basic concept of modularisation is graphically represented below.

As can be seen from Fig. 8.4, the approach to modularisation is divided into three essential steps. Initially, an existing product structure is analytically examined and broken down. Subsequently, crucial components are identified and selected. In a second step, the previously identified and selected components are analysed and examined for their suitability for modular grouping. Depending on the purpose of the modularisation, different evaluation scales can be used in the analysis. After the analysis is completed and essential components have been selected for their modular suitability, the third step takes place. In this step, the actual modules are defined and formed using selected components based on steps one and two, resulting in modules that are relatively sustainable in function and whose function can be individually checked.

A key characteristic of the modular approach is the concentration on functional couplings of individual components within the respective modules. Modules comprise components with complementary properties, which can thus be considered and tested as logical units. Figure 8.5 shows different views on modules according to Krause [83], supplemented by a sustainability view according to Koppenhagen [84].

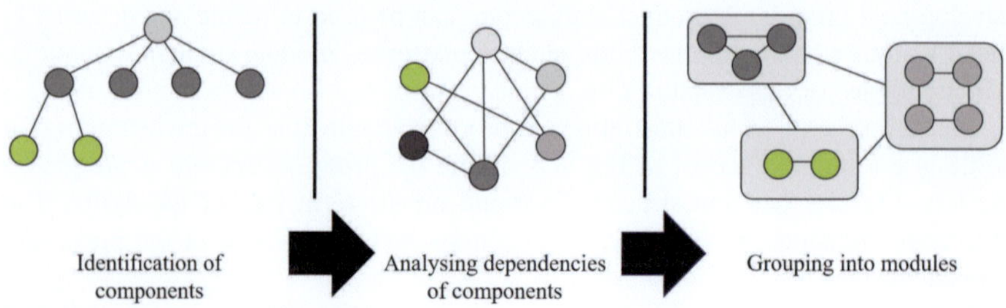

Fig. 8.4 Modularisation according to Krause [83]

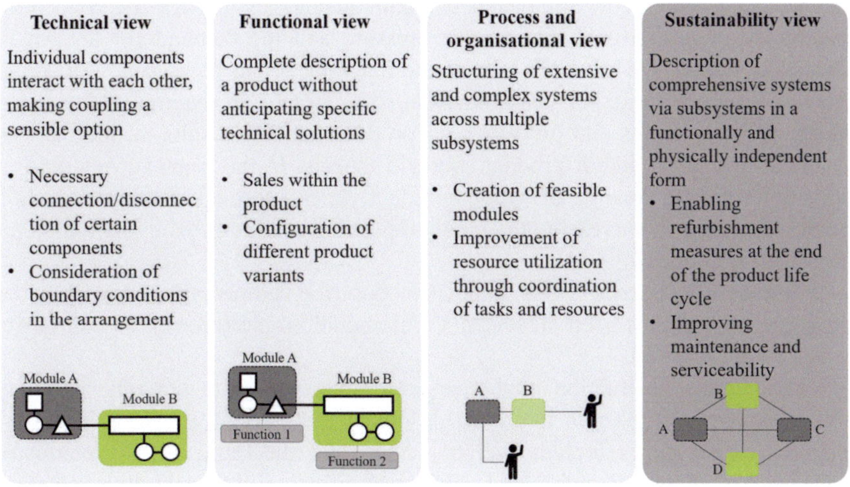

Fig. 8.5 Technical, functional, process, and organisational as well as sustainability view; own representation based on Krause et al. and Koppenhagen [83, 84]

From a technical perspective, it is important that the existing boundary conditions are taken into account when arranging the modules to be designed. In addition to this, Fig. 8.6 shows the relationships between production, costs, diversity, and improved competitiveness through the concept of modularisation, based on Rathnow [85].

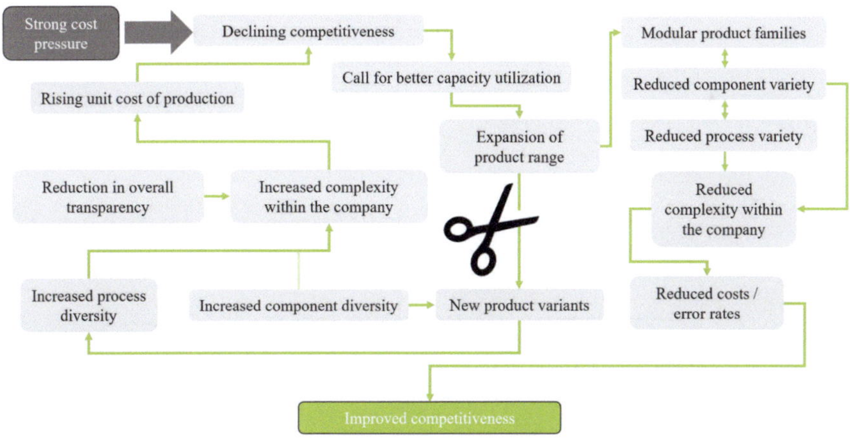

Fig. 8.6 Influences of modularisation on corporate competitiveness; own representation based on Rathnow [85]

The figure shows how companies can gain advantages through the concept of modularity in response to competitive pressure and the demand for an expansion of the range. On the right side of the diagram, it can be seen how modular product structure strategies can contribute to providing an externally demanded variety of components and processes for production. This results in interfaces to technical-functional and/or product strategic aspects. In this context, modularisation allows for the creation of various product variants through combinations in a simple way, thereby increasing the range of offerings and improving the long-term competitiveness of companies.

As previously described, this section will describe various types of designs. An extension of what has been presented so far should be the **integral or composite construction method**.

The integral construction method expresses the desire to combine as many individual parts as possible into a single component or to combine several subfunctions. Advantages include weight saving and the reduction of interfaces, connecting elements, assembly and adjustment efforts, and possibly the reduction of manufacturing resources for these processes.

The opposite is the composite construction method. It describes the breakdown of individual parts into several more favourable manufacturing parts. Components or even standard parts. In such a composite construction method, the assignment of individual functions to different components is enabled, thus promoting modularisation. Ultimately, the material selection for each part can be optimised according to load requirements. Repairs are facilitated, and only the functions that are needed are implemented. Figure 8.7 illustrates the relationships using the example of a housing unit.

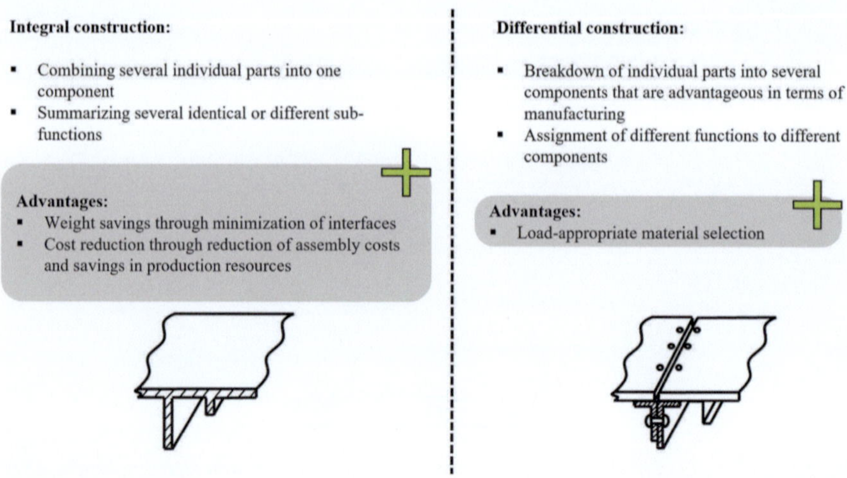

Fig. 8.7 Comparison of integral and composite construction methods using the example of a housing unit according to Krause et al. [82]

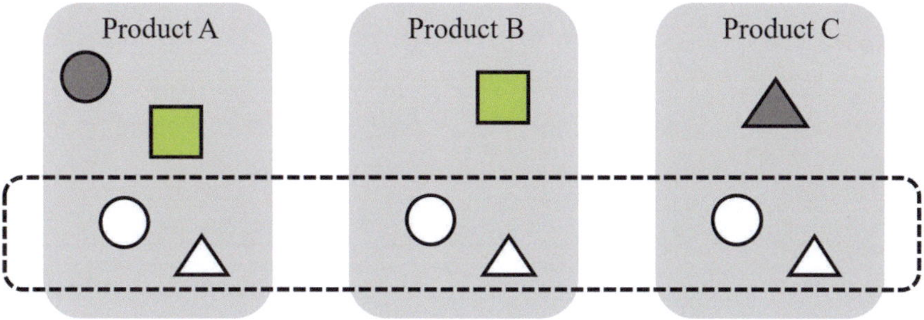

Fig. 8.8 Modular system and platform strategy; own representation based on Krause [83]

The efficiency of the two construction methods largely depends on the quantity of products produced. In general, it can be said that the composite construction method is more suitable for products with low quantities, while the integral construction method is ideal for products with high quantities. If considerations for potential repairs are neglected, highly integrated solutions are likely to be advantageous due to their lower number of parts and typically also due to the lower material usage. This is particularly true when considering ecological aspects. This applies at least in the case that no material composites or graded composite materials are used. Otherwise, the ecological impact needs to be reassessed.

Modular system or platform strategies describe a collection of various building blocks, which can be used to create products with different functions by combining selected individual modules. The modules are combined across all product variants of a product family and, optionally, also across several product generations. The following Fig. 8.8 schematically illustrates this relationship.

The relationship of the modular system or the platform shown in the figure is clarified by the consistently arranged structure of the components shown in white. These components are used for all three products (A-C) depicted and thus form the standardised platform of the product variants. This platform creates a common and overarching basis for the entire product family. By integrating further modules, which are shown in green and grey in the figure, different product variants can be developed subsequently.

8.2 9R Strategies

Product architecture and modularisation are essential enablers for a circular business model, which serves as a paradigm for a sustainable economy. Circular business models address, among other things, the avoidance of waste, the maximisation of usage intensity over several product life cycles, and the extension of the usage of materials in each cycle. To implement these circular principles, various strategies (such as recycling or refurbishment) have been developed. The

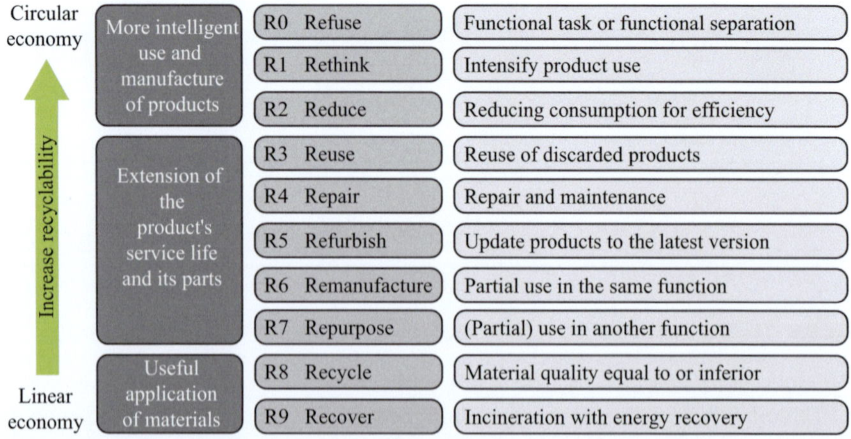

Fig. 8.9 R-framework; own representation based on Potting and Kirchherr [86, 87]

R-framework is a widely used method to define these strategies in detail and to illustrate them. The R-strategies derive their name from the initial letters of the respective strategies. The prefix "re" originates from Latin and stands for "new" or "again". In science, there is no uniform opinion on how many R-strategies can be assigned to the framework. Thus, the number of R-strategies varies between three and ten depending on the publication.

A common and frequently cited concept for understanding and representing the 9R framework comes from Potting and Kirchherr [86, 87]. For introduction and general illustration, this is shown in the following Fig. 8.9.

As can be seen from the figure, the framework arranges the strategies according to their degree of circularity. A higher degree of circularity in materials in a product chain indicates that these can be kept in the cycle longer and can preferably be reused with their original quality after the product's disposal. This reduces the need for natural resources for production. Accordingly, the R-framework illustrates that a higher prioritised R-strategy leads to lower environmental impacts in the production of a product, as the need for primary material for production decreases. However, this statement must be verified for each specific case. Although the need for primary material decreases, this decrease may be offset or even overcompensated by increased efforts in the return and reprocessing system.

An example of considerations within the R-framework in the context of recycling and resource conservation is beverage bottles. Water bottles are often filled at local sources, returned, and reused, representing a closed cycle in the sense of sustainable resource use. In contrast, wine bottles, which are filled with wine from global growing areas, are typically not intended for return. Instead, they are typically recycled by melting the glass and processing it into new bottles. This example illustrates the different approaches to resource use and circularity within the beverage industry.

In general, the 9R strategies can be divided into three guidelines: The strategies *R0 to R2* aim to avoid or reduce the raw material effort in production. This is achieved by making products redundant, as their functions can be provided elsewhere. Moreover, raw material consumption can be reduced through more efficient product manufacturing and usage. In this way, the same overall benefit for the customer can be ensured with fewer raw materials. The strategies *R3 to R7* aim to extend the lifespan of existing raw materials, which are already in the form of products within the economic system. Through the reuse or repurposing of various products or their parts, the benefit can be provided without additional or renewed raw material extraction. The last possible strategies are *R8* and *R9,* which are used when strategies *R0* to *R7* are no longer applicable. These serve to capture the raw materials of products or product parts that cannot be reused. The components are dismantled or recycled. By obtaining secondary raw materials, the extraction of new raw materials from the environment can be reduced.

The following list describes each of the R-strategies in detail and illustrates them with examples in the context of a kettle.

- **R0 Refuse**: The product or functions of the product are made redundant by abandoning functions or by offering different functions with a completely different product.

  A *new kettle can be avoided if there is no real need. In the case of a new purchase, a kettle is chosen that is reduced to the essentials and only contains functional elements.*

- **R1 Rethink**: Smarter/intensive use of products, development of more innovative and sustainable solutions.

  The kettle can be rented on a sharing platform for a limited time or provided in communal kitchens, thus being used more intensively and by more customers for a longer period. Other possibilities include the use of sustainable materials, the implementation of smart functions, or the use of more efficient energy supply options.

- **R2 Reduce**: Increasing efficiency in product manufacturing or usage by reducing the consumption of natural resources and materials.

  The use of recyclable materials reduces the need for new raw materials and extends the product's lifecycle. Alternatively, for example, heat recovery systems can store waste heat for later use, thus increasing energy efficiency.

- **R3 Reuse:** Reuse of a product that is still in good condition and fulfils its original function, by another user.

The kettle can be acquired in a second-hand shop, at a flea market, or on virtual marketplaces instead of being bought new.

- **R4 Repair**: Repair and maintenance of a faulty product so that it can be restored and continue to be used in its original function.

  The kettle can be repaired and reused by offering customer service or repair services, publishing repair instructions, or using a repair-friendly design. Long-term availability of spare parts can ensure the 'Repair' concept permanently.

- **R5 Refurbish**: Restoring an old/used product and bringing it up to date.

  The kettle can be professionally serviced and overhauled, for example, by the manufacturer. This includes a function check, repair of defective parts, and cleaning. Afterwards, the device can be put back into circulation and reused. Alternatively, new technologies can be integrated via services; for example, newer and more energy-efficient technologies can be installed and thus equipped for the long term.

- **R6 Remanufacture**: Use of parts from a discarded product in a new product with the same function. The integration of new functions is also possible here.

  Individual components of the kettle, such as heating elements, electronic components, or insulation parts that are still in good condition, can be recovered during disassembly and used in the production of new kettles.

- **R7 Repurpose**: The extended use of a decommissioned product or its parts/components in a new product with a different or altered function.

  The kettle can be used for other heating purposes, for example, in a larger water heating system. Alternatively, individual components such as heating elements or thermostats can be used and utilised in other electronic devices.

- **R8 Recycle**: Dismantling the product and processing materials for reuse to achieve the same (high-quality) or a lower (inferior) quality.

  When a kettle reaches the end of its lifespan, it can be recycled to separate and reuse the contained components. For example, metallic materials, plastic components, or glass elements can be recovered and reprocessed.

- **R9 Recover**: Maximising the benefit from products at the end of their life, e.g., incineration of materials with energy recovery.

In specialised waste incineration plants, unusable parts of the kettle can be burned, thereby recovering usable energy from these wastes.

Various value chains should be reconsidered and possibly restructured based on these 9R strategies to make them more resource-efficient. It is essential to integrate crucial and decisive decisions regarding the potential application of the 9R strategies already in the development phase. In particular, the previously explained concepts, such as product architectures, modular systems, and platforms, play a decisive role. These concepts must be used efficiently and optimised to ensure sustainable resource use.

Effects and Conceptual Design 9

The preceding chapters discussed specifications, evaluations, and general functions of products as well as their structures at an abstract level. The next steps towards the complete construction of a real product will now be outlined and discussed. The next two chapters of this book will show the path to very concrete designs and the final embodiment. The focus is therefore on how to progress from a rough and basic concept level via the conceptual design to an explicit realisation and component design in the development process. Chapter 9 will deal in detail with effects and designs, conceptual design principles, and the concept of EcoDesign.

The first section of this chapter begins with a general definition of effects and their significance for the discussed contexts. Subsequently, a focus will be placed on various transmission mechanisms, specific effects, their description and efficiencies, as well as associated functional variables. It will also deal with the assignment of effects to functions and sub-functions. The second section of this chapter introduces and defines the concept of the conceptual design. Essential contents will be the basic rules of conceptual design and general procedures. The third section of this chapter will list and describe conceptual design operators and their principles. Section 9.4 will then discuss various types of interference effects and emissions, to introduce the concept of EcoDesign and associated knowledge repositories in the last part of this chapter, and to put them in the context of sustainability-oriented design and shaping.

R. Lachmayer et al., *Methodology for the Development of Sustainable Products*,
https://doi.org/10.1007/978-3-662-72442-2_9

9.1 Effects

The desired overall function of a product can be described using general variables and operations. However, this does not yet provide a physical realisation. Based on the function structure (see Chap. 8), physical, biological, or chemical effects must therefore be determined that can realise the desired and required functions.

In general, effects can be described as a physical phenomenon that establishes a repeatable relationship between observed variables and characteristics of a definable system. In addition to physical effects, chemical and occasionally biological effects are also used in technical systems. A technical realisation of a function always requires an underlying effect. The number of usable effects is determined by the recognised laws of nature and is limited. Examples of physical effects are the lifting effect, the Coriolis force, the electro-optical effect or Hooke's law. Technical products can only function successfully if a corresponding natural effect is available for them to utilise.

Figure 9.1 shows a schematic representation of the effect and function size matrix according to Roth or VDI 2221. The matrix of effects and function variables consists of a tabular representation, where the input variables of the function are listed on the y-axis and the target variables on the x-axis. Each cell of the matrix then represents the relationship between a specific desired function and an effect. The effect sizes are physical quantities or parameters that influence the behaviour of the system (e.g., temperature, pressure, speed, etc.), while the functions describe the tasks or requirements that the system is supposed to fulfil (e.g., heat transfer, force transformation, etc.).

The functionality of the matrix is based on its ability to systematically collect the relevant functions and their impacts on the effect sizes. Users within product development can more easily select the right technologies, materials, and processes that best fulfil the functions of the system through this matrix. This is made possible by specifically identifying the effects that are necessary or beneficial for

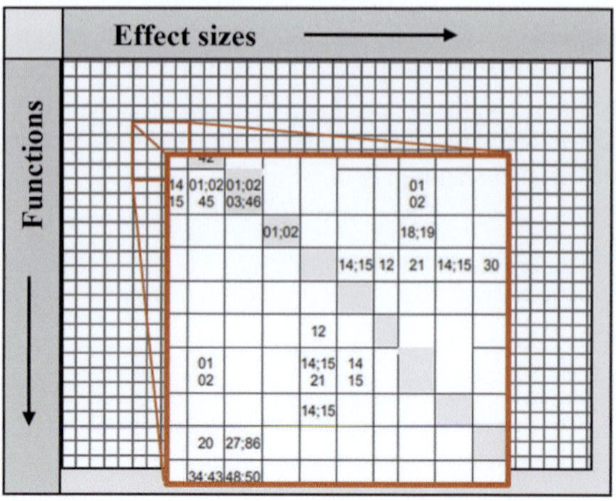

Fig. 9.1 Matrix of effects and function variables according to Roth [77]

achieving the respective function. In addition, interactions between different effect sizes and their impacts on multiple functions can be demonstrated, promoting a holistic view of the system in terms of potential synergies. The greatest benefit of the matrix of effects and function variables lies in its ability to optimise the design process by providing a clear overview of the relationships between the functions and the underlying technical effects. This advantage contributes to the reduction of complexity, enables efficient embodiment design, and prevents unnecessary redundancies in system development. Especially in the early stages of design, the matrix can help identify potential solutions and align the system with the most important requirements from the outset.

The matrix of effects and function variables finds its application in solution finding at an abstract level. The descriptions of the associated effects can be found in the inner matrix. Each of these effects is assigned a description, a principal sketch, a mathematical formula, and an example. In Roth's [77] publication, formulaic relationships and sketches relating to their conceptual design can be found for all effects. Empty fields indicate that no effects were found between the respective sizes on the x-axis and the y-axis.

To support the described theoretical context, various effects of representative significance for the context of a sustainability discussion are selected and discussed in the following section.

The **Peltier effect** is fundamentally described as a thermoelectric effect in which a conductor circuit, consisting of different metals, generates and dissipates heat at different soldered points. This effect always occurs when there is a current flow. This can lead to deformations [88] (Fig. 9.2).

With regard to the generation of technically usable energy, the **photoelectric effect** is worth mentioning. A solar panel, for example, can be used for the technical implementation of this effect. The function of the solar panel is to convert solar energy into electrical energy. The light intensity (I) that hits the panel directly influences the energy generated. The efficiency (η) of the panel indicates how much of the absorbed solar energy is converted into electrical power (P) (Fig. 9.3).

Another example is a heat exchanger that fulfils the function of **heat transfer** through the effects of convection and heat conduction. The effect sizes of the temperature difference (ΔT) and the heat flow (Q) can transfer the heat of one

Peltier effect	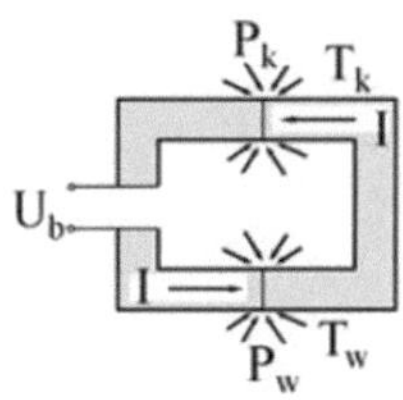	$Q = \pi I$
Heating or cooling at a phase boundary between two different electronic or ionized conductors when current flows (reversal of the thermal effect)		Cooling unit

Fig. 9.2 Peltier effect; figure according to Ponn and Lindemann [89]

Photoelectric effect	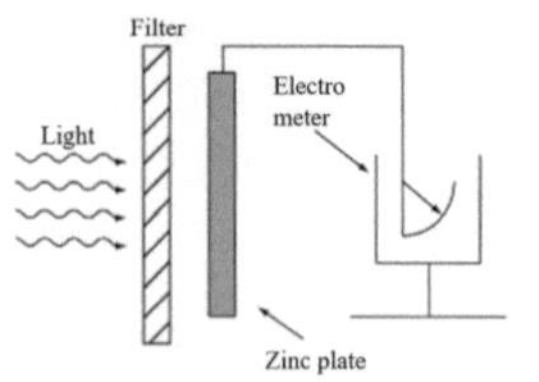	$E_{kin} = hf - W$
The photoelectric effect is the phenomenon in which electrons are released from a metal when it is irradiated with light of a certain frequency		Solar panel

Fig. 9.3 Photoelectric effect; figure according to Ponn and Lindemann [89]

Convection	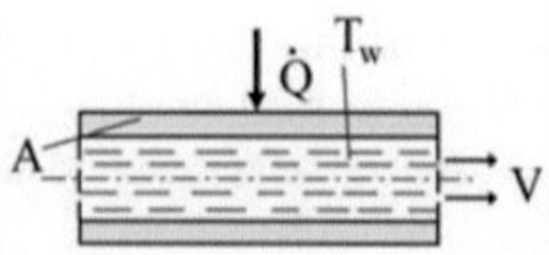	$\dot{Q} = h\,A(T_W - T_F)$
Heat transfer with simultaneous mass transport by free or forced flow of matter		Radiator, Heat exchanger
Heat conduction	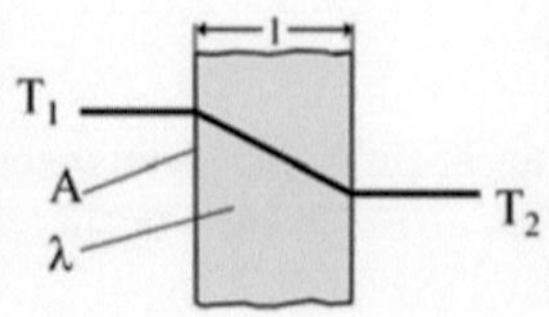	$\dot{Q} = \frac{\lambda A}{l}(T_1 - T_2)$
Heat transport inside a body or phase through lattice vibrations and moving charge particles		Heat exchanger, Insulation

Fig. 9.4 Heat transfer using the example of convection and heat conduction; figure according to Ponn and Lindemann [89]

medium to another. The temperature difference determines the efficiency of the heat transfer, the heat flow, and the amount of heat transferred (Fig. 9.4).

With regard to the function of generating usable energy, the **electric motor** is also worth mentioning in the context of a sustainability discussion. The function of this motor is to convert electrical energy into mechanical energy via the effects of induction. The converted energy is described by the voltage (U), current (I), flux density (B), and force (N). The effects described below are listed and presented in the known scheme (Fig. 9.5).

Unfortunately, the equations listed in the respective figures easily obscure the fact that the technical implementation of effects is usually associated with side effects and is negatively influenced by so-called disturbance variables. As a consequence, this means that efficiencies are effectively less than 100%. The control, compensation, or elimination of such disturbance variables requires considerable effort in some cases. To classify this circumstance, typical efficiencies of effects in technical implementation are listed in the following Table 9.1 as examples.

Induction (1)	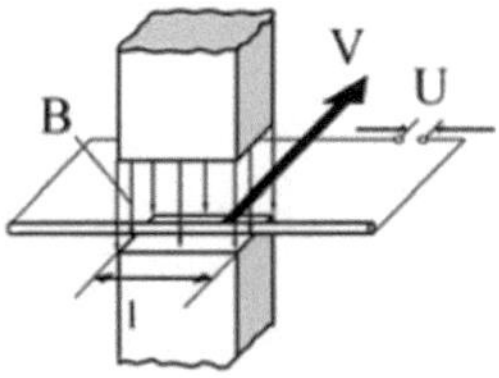 	$I = \frac{F}{B}\frac{1}{l}; \quad v = \frac{1}{Bl}U$
When the magnetic field changes over time through an open surface, a voltage is induced.		electric motor, speaker, moving-coil meter
Induction (2)	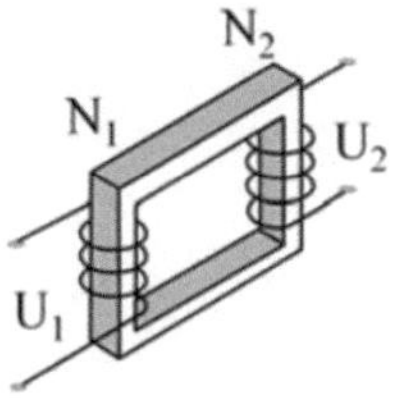 	$U_2 = \frac{N_2}{N_1}U_1; \quad I_2 = \frac{N_2}{N_1}I_1$
A voltage is induced in coil 2 by a magnetic flux generated by coil 1.		Transformer Transmitter

Fig. 9.5 Electric motor using the example of the effect of induction; figure according to Ponn and Lindemann [89]

Table 9.1 Efficiencies of selected effects in technical implementation

Effect	Technical implementation	Efficiency (%)
Photoelectric effect	Solar cell	20
	LED	30
Force transformation	Lever	99
	Pulley system	95
Thermoelectric effect	Peltier element	5
	Seebeck effect generator	3–8
Electromagnetic induction	Generator	98
	Transformer	< 50–99
Heat transformation	Heat pump	45–55
	Refrigeration unit	24–52
Sound wave conversion	Microphone	< 1
	Speaker	0.2–2

It is fundamentally important to note that effects must be discovered. Consequently, even with high effort, solutions cannot be forced. However, there is the possibility that existing efficiencies can be improved to a certain extent through technical effort. This usually involves a cost–benefit analysis of the effort for the technical implementation versus the benefit to be realised.

Material elements are required for the implementation of an effect. These are referred to as effect carriers. Effect carriers always rely on the specification of geometric sizes and a material. Considering the geometric elements of an effect carrier, working paths, active surfaces, and active spaces can be distinguished.

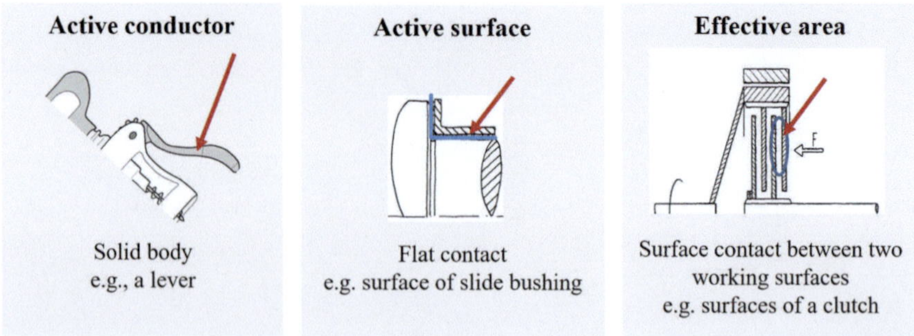

Fig. 9.6 Types of effect transmission

The transmission behaviour of effects within a system or between components of a system is directly dependent on the design of these components. The embodiment design of a pair of active surfaces for connecting different components can thus be crucial for the successful transmission of an effect. A typical example of this is bearings in an electric motor, which represent a common pair of active surfaces, or the shaft of such a motor, which acts as a classic working path.

Figure 9.6 schematically illustrates the transmission of effects based on working paths, active surfaces, and active spaces.

A linear effect is achieved through a working path, as is the case with supports or levers, for example. In the case of surface effects, such as the contact of two tooth flanks, a pair of active surfaces is usually formed. A working space results from a volume function, such as the dielectric of a capacitor, or from the space-creating surface contact between two coupling surfaces.

The question arises as to how these form-function relationships in a system can be specifically described and influenced. The functions and thus the necessary physical effects have their consequence in the design of the embodiment. The possibilities for a successful approach are listed below:

- **Geometric Scaling**: In the case of extremes in the size of systems (both extremely large and extremely small systems), it is advisable to carry out manageable scaling. The abstraction to a system size that is known allows prototypes to be developed in product development and then adapted to the desired size extremes.
- **Reconstruction**: Reconstruction is used in product development to replicate functions and sub-functions using known components to gain a better understanding of the form-function relationships. For example, by disassembling entire systems, existing pairs of active surfaces can be identified, and the knowledge gained can be used for further prototypes.

Simulation: Simulations can be used to test and evaluate the suitability and potential success of identified and validated parameters.

9.2 Conceptual Design

The purpose of the conceptual design is to move from the level of effects to a more concrete design. In the process of conceptual and embodiment design, the design features discussed in more detail in Chap. 10 are gradually concretised. The conceptual design initially allows for a clear representation, variation, and optimisation of the overall relationships. The following four basic rules should always be observed when designing:

- **Clear**: For a clear construction, clear effects must be present, a clear interpretation must be made, and the design must be clearly defined. The assembly sequence must be clear, and materials must be clearly separable for recycling.
- **Simple:** In a simple construction, only the necessary sub-functions are present. Parts are designed simply and, taking into account basic geometric functions and symmetries, have a simple embodiment. Tolerances are only as fine as necessary.
- **Safe**: In a safe construction, all hazards are eliminated or minimised, necessary protective measures are considered, and users are informed about residual risks.
- **Sustainable**: This aspect includes the development of efficient products, saving resources, and economic operation. The tension between the ecological, economic, and socio-cultural impacts of a product must be analysed, optimised, and balanced throughout the product life cycle.

Archetypes for the conceptual design process can be described as follows:

Type A: Based on working paths, working surfaces, and working spaces
In this case, these aspects are derived from the effects and the function structure and expanded by dimensioning the design sizes around the material-dependent necessary spaces. In addition to the effect geometry and the arrangement of effect surfaces as well as effect lines and spaces, the geometry is completed by taking into account, for example, manufacturing aspects.

Type B: From inside to outside
Starting from the performance data, the components inside a machine are first dimensioned. The construction of the machine around these components then follows to achieve the required dimensions. Examples are electric or combustion engines, pumps and turbines, or gearboxes. In the design of gearboxes, axle distances are defined based on the desired performance of the engine, torques, and speeds, and shafts and gears are dimensioned. The requirement-appropriate design of the interior consequently leads to the adaptation of the exterior. As a result, the gearbox housing is adapted to the performance of the engine as a superior system, not the other way around.

The following Fig. 9.7 shows an example of a gearbox to illustrate the above, which is used in a practical teaching event, among other things, to demonstrate the relevance of this archetype to first-semester engineering students.

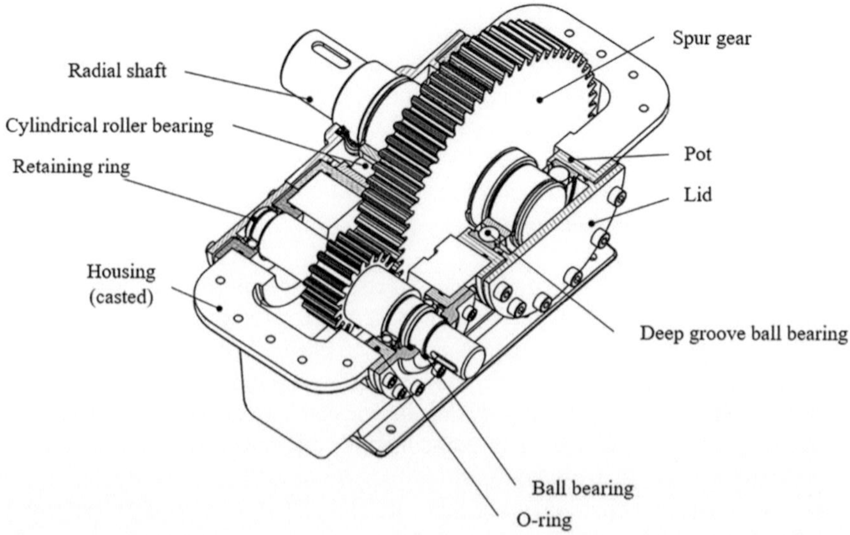

Fig. 9.7 Representation of a gearbox

Type C: Based on standard components

Components of a system for the specific application are combined from catalogues. Examples are energy, electrical systems, heating systems, or logistics systems. The design of combined heat and power plants is based on an estimate of the real application. Depending on these assumptions, standard components such as mass-produced engines are combined into an overall system and offered to the market. As a result, systems are created that are combined for a specific application case by linking components available in modular systems. For example, numerous standard components in the form of switch boxes, distribution systems, plugs, and cable systems of different performance classes are used in the context of electrical installations and distribution systems. This not only saves costs but also ensures that the systems work reliably and efficiently through standardised, tested components. Detailed geometries often do not have to be elaborately developed and documented in technical drawings, but are merely integrated with their connection dimensions into so-called circuit, flow, or piping plans and the associated parts lists.

Type D: From outside to inside

Starting from cabinet, housing, or container sizes, systems are designed to fit into these spaces. Examples are computers and computing systems, but also the interior design of mobility systems. The expensive infrastructure with the associated standardisation must be taken into account. The example of Deutsche Bahn's wagon formations (see Fig. 9.8) shows that the outer boundary of the system is clear. Based on these dimensions, it is necessary to design the carriages of the vehicle

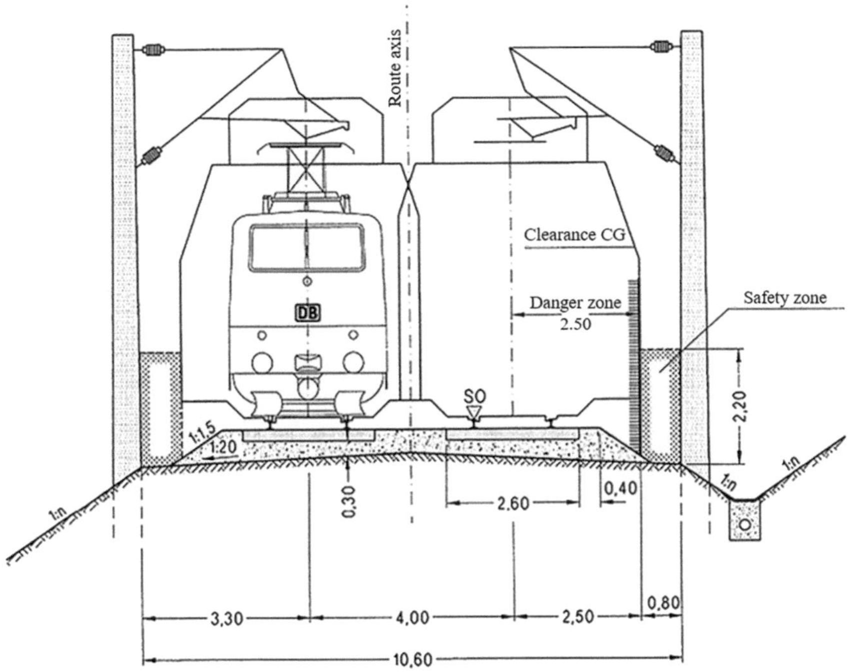

Fig. 9.8 Representation of a rail vehicle in the railway system [90]

and its interior. In doing so, the energy and information technology infrastructure must also be taken into account.

Type E: Through Symmetry

In the context of the conceptual design and manufacturing, symmetry is often deliberately used to achieve specific advantages. For example, the left and right sides of a product are often mirror-symmetrically designed, while certain arrangements exhibit rotational symmetries. Moreover, entire products exist in mirror-image versions, such as for left- and right-hand traffic or for use by left- or right-handed individuals.

Type F: Through Scaling

Scaling is a proven method of development for constructing series. In this process, lengths scale linearly, areas scale with length and volume, and weight scales in the third order with lengths. However, functional aspects and sizes lead the product scaling, which is why scaling is often done based on arithmetic or geometric series, or even more complexly, for example, in fluid mechanics, based on similarity numbers such as the Prandtl, Nusselt, or Reynolds number. In DIN 323, decimal geometric series are used to determine standard numbers. For the R5 series, these are, for example, 1, 1.6, 2.5, 4.0, 6.3, and for the R10 series, 1, 1.25, 1.6, 2.0,

2.5, 3.15, 4.0, 5.0, 6.3, 8.0. These can be found in many products, such as the diameters of drills or the widths across flats, in plugs, cross-sections of cables, or pipe diameters.

Side Effects

In addition to fulfilling intended functions and taking into account the previously specified boundary conditions, the conceptual design must increasingly consider disturbing (side) effects, also referred to as interference effects in this case. These disturbing side effects are functional, manufacturing, or operational effects or impacts that necessarily occur as side and secondary effects during the realisation of the desired effects, during the manufacturing or operation of a product. Examples include deflections due to self-weight, heat loss when converting energy, exhaust gases, wear and tear, and abrasion of microparticles, or even corrosion. The need to control these interference effects can significantly complicate designs, contrary to the guiding principle of simplicity mentioned above.

Losses usually mean energy dissipation through heat. Especially at low efficiencies, this must be constructively considered to avoid excessive heating and/or thermal deformation of structures. If heat is generated by friction losses, this unwanted side effect can usually be reduced by lubrication. However, this implies more or less complex constructive measures for sealing grease or oil as well as for filling and cleaning.

Other frequently occurring and unwanted side or interference effects are vibrations caused by the working process of a machine, geometric inaccuracies of the components, or external influences. These can lead to high component loads or unwanted noise development. To reduce vibrations, various constructive measures are available, such as shifting excitation frequencies, inserting dampers or snubbers, or changing the natural frequencies by redesigning.

In the conceptual design, side effects can be divided into two manifestations: On the one hand, they occur from the inside out and result from manufacturing and material inaccuracies as well as the operation itself. Exemplary outwardly directed side effects are thermal deformations and expansions, form and position deviations, and vibrations occurring inside the component. On the other hand, side effects also occur from the outside in. Examples here can include temperature fluctuations, humidity, and resulting corrosion or incoming dirt. These interference effects must also be appropriately considered in the conceptual design.

In general, some guidelines can be formulated for efficient handling of effects in design:

- The **energy conversion** is lossy and should be avoided where possible. Ideally, one should work with one form of energy in a system to keep losses and complexity low.
- The **transformation** of **energy and material** is also associated with losses and should therefore be reduced to the necessary level for the system.
- The **efficiency of storage** is highly dependent on the form of energy and the duration of storage. Solids and liquids can be stored better than gases, but these

are still better than electrochemical reactions (energy in batteries). The higher the energy content per volume, the smaller the storage.

- Processes such as **conducting**, **linking**, and **separating** are usually also associated with losses. The transportation of energy (work), power, or forces is inefficient and leads to material waste, with energy being easier to handle than power.

In addition to the consumption-oriented energy consideration, the holistic analysis of efficiencies and ecological impacts shows that the loss of physical resources of a product—whether due to high material wear during usage or due to a small proportion of reusable parts at the end of the product life—influences the eco-efficiency significantly over the entire life cycle.

9.3 Conceptual Design Principles

The design to be developed represents the sum or the summary of all design features necessary for the clear description of a component, an assembly, or an entire system. Design features are, on the one hand, concrete properties and, on the other hand, descriptive parameters of the embodiment. While design features can be directly influenced in product development, the properties of the design are resultant and dependent on the features. For example, the resulting weight of a component, which is to be optimised in terms of resource expenditure, is significantly influenced by its dimensions and the material used. A design is only considered efficient when as many of the design features as possible have a direct influence on the function of a component. In the development of resource-efficient and function-oriented products, the causal relationship to be considered in the conceptual design is multifaceted and complex.

An exemplary approach to the systematic application of conceptual design operators is, for example, based on the SCAMPER method by Eberle [91]. Along with the acronym consisting of "**S** ubstitute, **C** ombine, **A** dapt, **M** odify/ **M** agnify, **P** ut to other use, **E** liminate and **R** earrange/ **R** everse", new variants of the design can be identified starting from a reference component. The implementation of the SCAMPER method or comparable methods is initially based on the presentation of the reference component and the associated challenges. These are transformed into corresponding questions and answered in teams in the form of concrete solution variants using the design operators. Table 9.2 provides an overview of the manifestations of these conceptual design operators of the described approach.

This approach leads to the idea of product generation development, mainly developed by Albers [92]. Products are based on predecessor generations that pass on essential knowledge to the development of the next product generations in the sense of technical inheritance (see Chap. 4). From one product generation to the next, only the necessary innovations are introduced or developed that are necessary to reduce the development risk and minimise development efforts.

Table 9.2 Conceptual design operators according to the SCAMPER method by Eberle [91]

Conceptual design operators	Manifestations
Substitute	Functions, components, materials, energy, …
Combine	Functions, effects, applications, components, …
Adapt	Applications, components, functions, effects, …
Modify	Ergonomics, properties, functionality, embodiment, …
Put to other use	Market, users, application, purpose, …
Eliminate	Functions, components, effects, materials, …
Rearrange	Embodiment, components, production processes, …

At the functional level of a product, the targeted use of conceptual design principles allows for the manipulation of physical effects and an experience-based, heuristic approach. The principles listed below must be specifically weighted and prioritised, as without the use of artificial intelligence, optimisation calculations, and computer-aided implementation, they do not represent a closed and consistent system. Instead, they qualitatively describe different experiences without producing a clear solution. Principles for conceptual designing can be exemplified as follows:

Principle of direct and short force transmission

This principle aims to avoid unnecessarily "wandering" forces. With minimal deformation of the component, forces should be directed purposefully to the force discharge point. This approach can minimise material use and design a lightweight design. The topology of the component is a significant embodiment design characteristic at this point. The stiffness-mass ratio, for example, shows concrete effects of structural optimisations regarding the necessary material and resulting potential for lightweight applications. Particularly critical and to be avoided are constructions that lead to large bending moments and thus stresses on the structure. Here, symmetry (see Sect. 9.2, type E) is a suitable means to improve the conceptual design. The principle is schematically illustrated (Fig. 9.9).

In the figure, all three elements experience the same load. The effects of force transmission are compared depending on the design of the geometry. On the far left, the indirect force transmission is shown, in the middle, the indirect but closed force transmission, and on the right, the direct force transmission. It can be seen that the part depicted on the left requires more material than the one on the right to withstand the same loads.

The principle of uniform structural strength

The material stress in a part should be evenly distributed to avoid the occurrence of stress concentrations that can lead to premature failures. Stress peaks should be avoided as they can cause material fatigue and ultimately the failure of components. Special care should be taken with notches and corners during construction, as these areas are prone to crack formation and further damage. The principle is

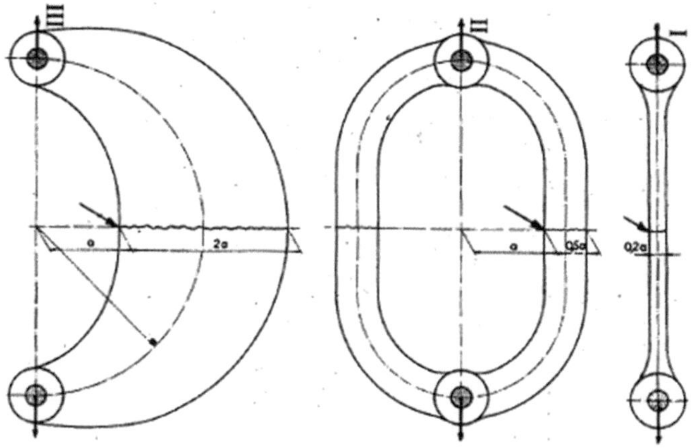

Fig. 9.9 Principle of direct and short force transmission; figure after Ehrlenspiel [93]

illustrated below, and constructive countermeasures to harmonise *force flow lines* are shown in the form of relief notches (Fig. 9.10).

The principle of coordinated deformations

Coordinated deformations or the “release” of deformations are sometimes also decisive for the functionality of the construction. The rotation of gear shafts and gears can, for example, lead to constraints in planetary gears. In the case of sliding bearings, a deformation coordination is necessary to avoid edge carrying. Figure 9.11 illustrates the principle, with the values F standing for acting forces and τ for resulting stresses.

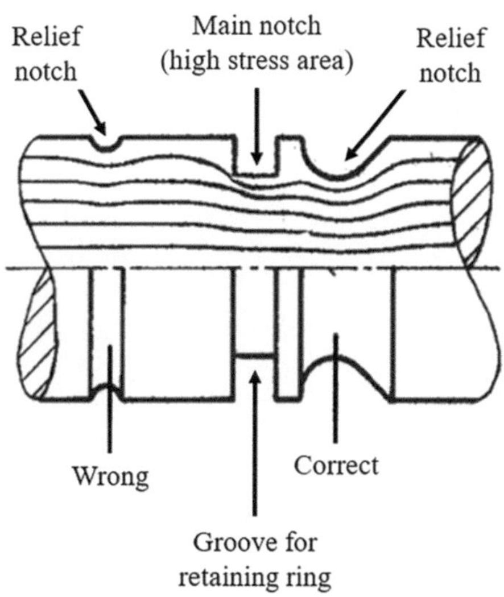

Fig. 9.10 Principle of uniform structural strength; figure after Ehrlenspiel [93]

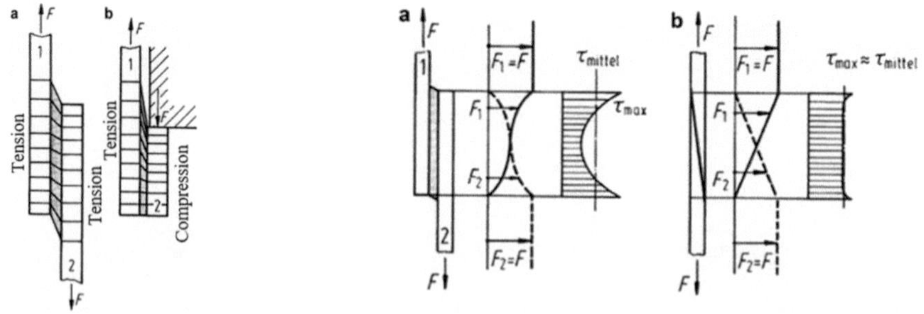

Fig. 9.11 Design principle of coordinated deformations according to Pahl and Beitz [94]

The principle of force balance

When considering the principle, two types of forces are assumed to be given. These are forces which serve the immediate fulfilment of the function (function-related main sizes), and on the other hand, these are consequential forces which result from the action of the main sizes and cannot be avoided (accompanying secondary sizes). The aim of the force balance addresses the necessary compensation of such secondary sizes. This can be achieved either by symmetrical arrangements of structural elements or by providing compensation elements. Figure 9.12 illustrates the principle using the example of a crane. While lifting the mass, m_L represents the main function, resulting secondary sizes, such as the bending forces acting on the mast, must be compensated for by attaching a counterweight. When moving the masses, synchronised movements must take place.

Principle of function-fulfilling stiffness

Depending on the development goal, constructions should be particularly stiff or have defined areas of elastic deformation. These can be set during the design process, as shown in Fig. 9.13, using the example of the embodiment design of a spring. In the example listed, E stands for the spring element, b for the ring width, D for the inner or outer diameter, γ for the cone angle, L for the length, s for the spring travel, and t for the safety gap.

The principle of error-redundant design

The principle aims to improve the reliability and security of systems through the use of redundancies, error detection, and correction. This includes, among other things, the installation of duplicate components with the same function and mechanisms for automatic error detection, compensation, and correction, to prevent or rectify system failures. The design is also robust, so that a variety of error conditions are tolerated without leading to a complete failure. This approach is particularly important in critical areas such as aviation or medical technology, to ensure continuous functionality.

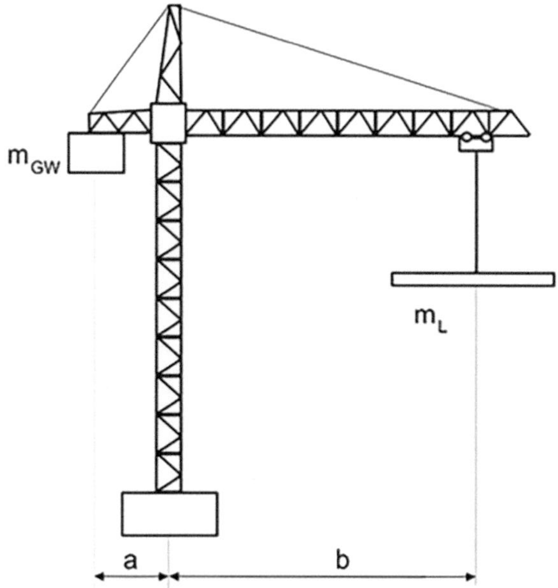

Fig. 9.12 Schematic representation of the design principle of force balance

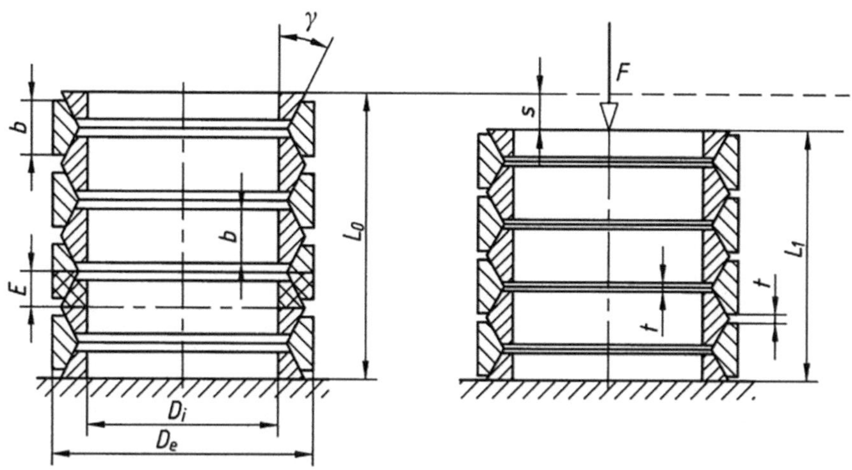

Fig. 9.13 Principle of function-fulfilling stiffness using the example of a spring; figure after Spura et al. [95]

Principle of self-protecting solution
The purpose of the principle is to prevent major damage or destruction in case of overload by introducing a secondary effect. This is usually obtained from an additional force transmission path, which is initiated in case of overload. This results in a different force flow distribution and consequently a reduced load. However, this change often also results in changes to the functional properties, which can be either reversible or irreversible, as in the case of an emergency break point.

Principle of Stability/Bistability
When considering the conceptual design principle of stability or bistability, the ability of a system is examined to exist in one or two stable states. A bistable system can return to one of two possible stable states after an impact, depending on which state is favoured by the input or environmental influences. An example of this is a conventional light switch. In design, stability means that a system is designed in such a way that it remains stable even when exposed to external forces or changes. Stability or bistability can be achieved through mechanical, electronic, or other design elements that keep the system in a stable state or allow a transition between two stable states (Fig. 9.14).

The figure shows that a ship does not capsize and remains stable as long as its centre of gravity S is below its pivot point M. In the ship depicted on the right, the centre of gravity is too high, and it will capsize as soon as the centre of gravity moves out of the vertical axis of the pivot point M due to disturbances. Between stable and unstable system states, there are also situations in which systems, for example, after a disturbance, transition into indifferent states. These should be consciously avoided in the design.

Principle of self-reinforcing solution
In the principle of the self-reinforcing solution, for example, the combination of friction and adhesive force is exploited to amplify clamping forces. The same principle is used in the safety devices of elevators. Seals are also often arranged

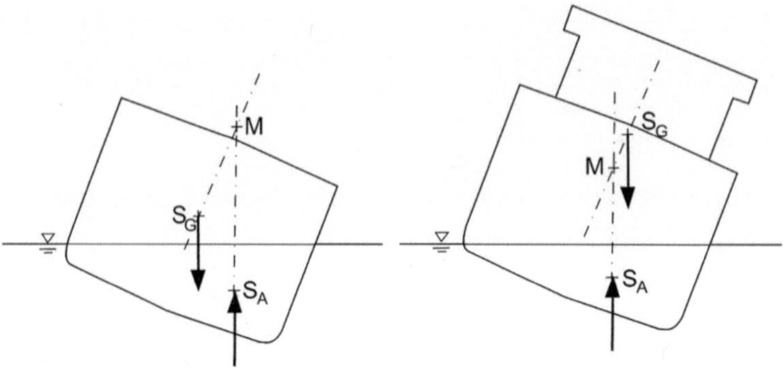

Fig. 9.14 Schematic representation of the design principle of stability/bistability

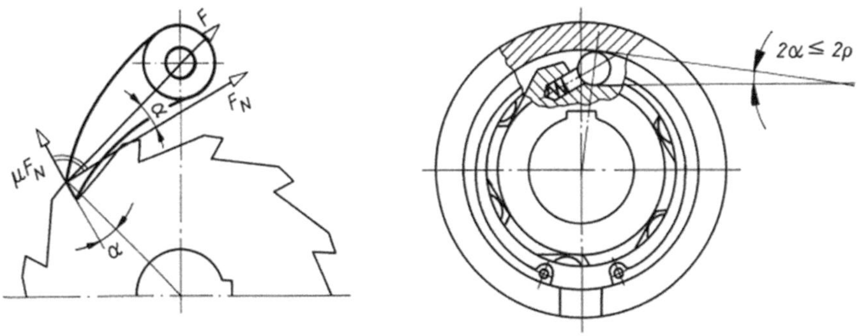

Fig. 9.15 Principle of the self-reinforcing solution using the example of a freewheel clutch; figure after Spura et al. [95]

so that they are pressed harder with increasing internal pressure, thus increasing their effect. Figure 9.15 shows a free wheel, which also operates according to this principle and locks more strongly the harder it is worked against. Here, F stands for the pawl force, which is at an angle α to the normal force F_N. Furthermore, in this example, the coefficient of friction is represented by μ.

Further examples can be found in standard works such as Pahl/Beitz [96] or Roloff/Matek [95] in various editions.

9.4 Emissions

Even in the conceptual design phase, it is necessary to deal with the emissions of a future product. Emissions can occur as gases, liquids, or solids. Emissions can also be noise, light, or other forms of radiation. Depending on the potential hazard, these elements should be distributed deliberately and with a low threshold, with the observance of distances, the provision of enclosures, and the careful avoidance of emissions being essential. In addition, the targeted collection, storage, and proper disposal of the relevant materials or substances are often provided. In relation to the conceptual design, this means a targeted examination of the necessary measures, their consideration, and often a significant complication of the system (Table 9.3).

The following will further illuminate the example of wind energy emissions. Wind turbines generate aerodynamic noise through their rotating blades, which depend on the blade tip speeds and therefore lead to increasingly critical aerodynamic measures in the design of the blades as rotor diameters increase. Mechanical components that also contribute to the noise development of wind turbines include the gearbox, the generator, the fans, and the auxiliary drives for adjusting the system. These mechanical noise sources have, in the past, led to frequency-dependent,

Table 9.3 Exemplary manifestations of emissions in air, water, and soil

Type of emission	Air	Water	Soil
Gases	Carbon dioxide (CO_2), Nitrogen oxides (NO_x), Methane (CH_4)	Ammonium (NH_4^+) (dissolved in water), dissolved gases such as CO_2	Fluorocarbons (at landfills)
Liquids	Aerosols (e.g., spray mist)	Oil spills, chemicals such as pesticides	Chemical leaks from containers
Solids	Fine dust (PM2.5, PM10), soot particles	Sediments, microplastics	Heavy metals, plastic waste
Noise	Traffic noise, aircraft noise, industrial noise, Wind turbines	Water traffic noise, noise pollution in ports, and offshore construction sites	Machinery noise in agriculture/forestry
Light	Light pollution from street lighting, illuminated signs	Underwater lighting (at port facilities)	Light pollution from industrial plants
Radiation	UV radiation, electromagnetic radiation (e.g., from mobile phones)	Radioactive substances (e.g., released into water by nuclear tests)	Radioactive deposits

disturbing single tones. However, such noises have been significantly reduced in recent years. For newer types of systems, there are also detailed regulations for recording and limiting specific noise emissions and for maintaining minimum distances from buildings. To put this into context, Fig. 9.16 shows the noise emissions of a wind turbine compared to other noise sources.

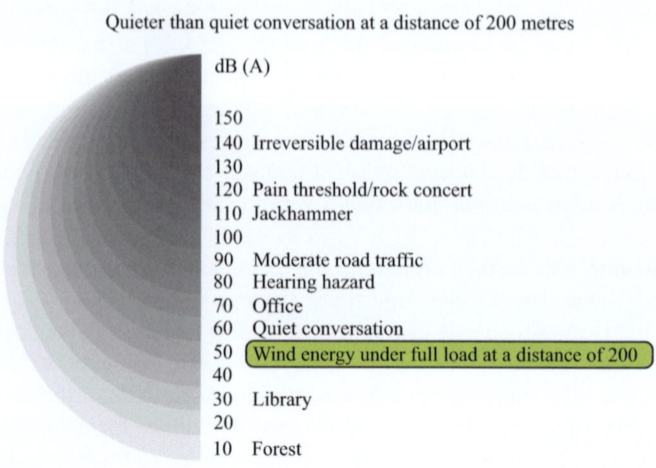

Fig. 9.16 Noise emissions of a wind turbine in comparison; own representation based on BWE [97]

In addition to these noise emissions, other emissions from wind turbines need to be described. For example, such systems have obstacle lighting to protect air traffic at certain heights. These must be switched on as required, i.e., only when aircraft approach, resulting in light emissions. Currently, the solid abrasion of the blades is also being discussed as it can have effects on the environment and the longevity of the systems.

Another example, which will be explained in more detail below, deals with sewage treatment plants in Germany and the resulting emissions. Approximately 1.6 million tonnes of dry mass are produced annually in the approximately 10,000 wastewater treatment plants (sewage treatment plants) in Germany. This is created by drying sewage sludge, which is either done using solar heat or in industrial drying plants where moisture is specifically removed from the sewage sludge. This is usually done using fossil or renewable energy, or possibly waste heat from other processes. According to new federal legislation, the dry mass from sewage treatment plants can no longer be spread on fields as fertilizer due to its contamination with various pollutants, but must be thermally recycled or used as fuel. Before this recycling, usable phosphorus is extracted from the fertilizer.

9.5 EcoDesign

The principle of EcoDesign can be defined in a general context as a practice where holistic ecological, economic, and technical optimisations are integrated into the product design process. A common definition is that EcoDesign includes sustainable solutions that can be characterised as products, services, hybrids, or system changes. These solutions, through their specific design, are capable of minimising negative impacts on sustainability, both ecologically and economically, as well as socially and ethically. At the same time, it is possible to fulfil societal requirements in this way [98].

A proven and practical method in this context is, for example, the approach according to Design for X (DfX) strategies. The principle of EcoDesign can itself be referred to as a DfX process, where the X can stand for various life cycle phases, product features, or aspects of design. It should be noted that the claim of EcoDesign includes a holistic consideration of the entire product life cycle, which distinguishes it from other, often specifically pronounced DfX approaches [99].

A tool for concretising and transferring EcoDesign into actual product development has been elaborated and proposed by Luttrop and Lagerstedt [100]. They define a set of rules ("the ten golden rules") in the described context, with the help of which developers and designers are to be supported in the inclusion of sustainability aspects in product development. The set of rules provides a kind of guideline for the conceptual design of situation-specific product designs and for overcoming resulting challenges. The ten rules are summarised and listed in the following list.

1. Avoid toxic substances and use closed cycles for necessary but toxic substances.
2. Minimise energy and resource consumption in the production phase and during transport through improved housekeeping.
3. Use structural features and high-quality materials to minimise the weight of products, provided this does not affect the required flexibility, impact resistance, or other functional priorities.
4. Minimise energy and resource consumption in the usage phase, especially for products with the most important aspects in the usage phase.
5. Promote repairs and upgrades, especially for system-dependent products such as mobile phones, computers, and CD players.
6. Promote the longest possible lifespan, especially for products with significant environmental aspects outside the usage phase.
7. Invest in better materials, surface treatments, or structural precautions to protect products from dirt, corrosion, and wear and tear, thus ensuring lower maintenance effort and a longer product lifespan.
8. Prepare for upgrades, repairs, and recycling through access options, labelling, modules, predetermined breaking points, and manuals.
9. Promote modernisation, repair, and recycling of products by using fewer, simpler, recycled, unmixed materials and no alloys.
10. Use as few fasteners as possible when connecting components and use screws, adhesives, welds, snap connections, geometric locks, etc., according to the life cycle scenario [100].

In the context of the design level and the initial component design, the aspects mentioned in the list can be considered helpful and supportive. In the nomenclature of this book, however, they are subject to the same restrictions as the previously mentioned principles or guidelines for design, in that they are neither completely free of contradictions nor necessarily lead to clear solutions, but can only serve as suggestions.

In addition to this detailed support and implementation approach, there are a large number of other tools, guidelines, and other aids under the term EcoDesign. Table 9.4 lists some further examples of selected tools of this type. The listing is done according to the different phases of the development process in reference to McAloone and Pigosso [99].

It should be noted in this context, however, that this is not a closed system, but various individual aspects described by different influences and sometimes contradictory terms. Various definitions and associated system boundaries must be merged to successfully implement the principle of EcoDesign in conceptual design and embodiment design.

In this area, there is both a significant scientific challenge and an urgent need for action to arrive at pragmatic and applicable approaches and tools.

Table 9.4 Tools for integrating EcoDesign into various development phases according to McAloone and Pigosso [99]

Phase	Tool/instrument
Product planning, task clarification	STRETCH-tool
	C2P-platform
	Eco-roadmap
	Eco-QFD
Concept phase	MECO-matrix
	DfE-Matrix
	Alternative function fulfillment
	Eco-function matrix
Design phase	Ecodesign-PILOT
	Fast five awareness tool
	Life cycle assessment (LCA)
Elaboration phase	Ecoquest
	Eco-communication matrix

Embodiment Design and Technologies

10

After the effects and conceptual designs were discussed in detail in the previous chapter, this chapter will explain how sustainable products can be designed in concrete terms. This will cover aspects of the specific embodiment design, from lightweight construction principles to exemplary technologies, manufacturing processes, and materials. The following is based on the definition that embodiment design is understood as the sum of all design features relevant for industrial manufacturing. Starting from specification, concept, and conceptual design, the embodiment design of a product is closely linked to the technologies, materials, assembly techniques, intended functions, and stresses envisaged for implementation.

In the following sections, in addition to general explanations on embodiment design, three examples of technologies and materials in the context of a sustainability-oriented embodiment design will be discussed. The technologies and materials examined in this context are:

- Additive Manufacturing,
- Bio-based materials and composites,
- Wooden constructions.

10.1 Embodiment Design

As already described in Chap. 9, the process of designing, dimensioning, and detailing involves the successive determination of numerous aspects of the product architecture and all associated components. The term embodiment design or the activity of (embodiment) designing also comprehensively includes all activities that determine the shaping or embodiment of products. This also means that the

R. Lachmayer et al., *Methodology for the Development of Sustainable Products*,
https://doi.org/10.1007/978-3-662-72442-2_10

design features are defined down to the smallest detail. Design features in this context are those features that define the embodiment of individual parts, assemblies, and systems.

Figure 10.1 illustrates that the design features of an object are described both by geometric properties—characterised by topology, shape, dimensions, and number, as well as tolerance specifications—and by the material properties and surface properties. Depending on the complexity, the expression of these features can vary in a part or within a product and occur at different frequencies.

In the following Fig. 10.2, selected design features are illustrated and presented using the example of a steering wheel.

The phase of embodiment design is therefore characterised by the transition from the rather generalist design level to the very concrete function- and production-oriented realisation.

In order to enable the best possible embodiment design in the process of shaping, various approaches described in detail in the literature can be consulted. All of these approaches have in common that the design process is designed in such a way that the final embodiment of the technical product is supported suitably by a systematic approach. The following are some examples of these approaches and their descriptions:

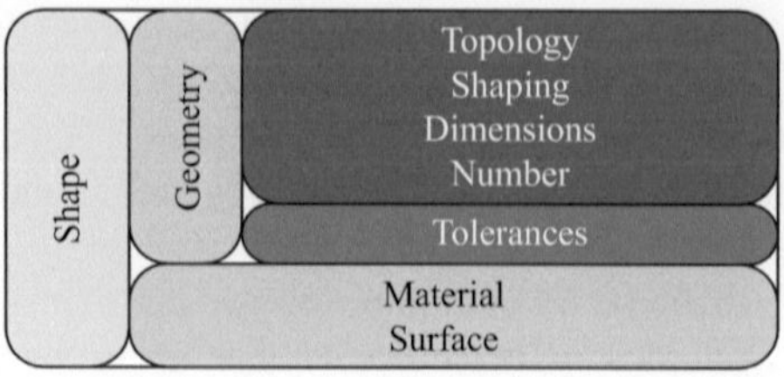

Fig. 10.1 Systematics of design features

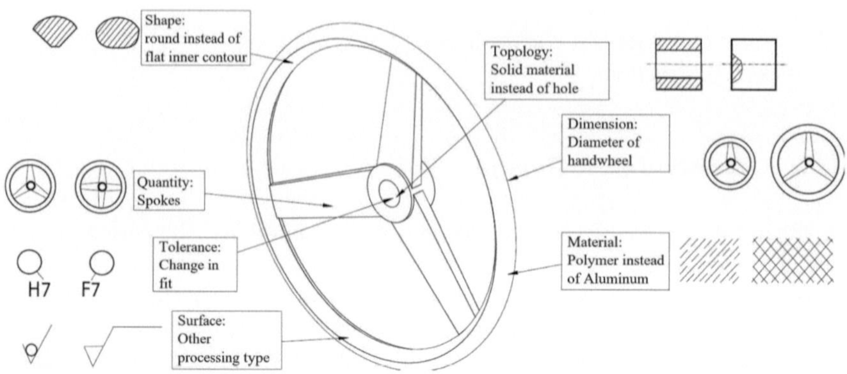

Fig. 10.2 Design features using the example of a steering wheel

Combination by Design of Elementary Structural Elements

This approach, formulated by Roth [35], is based on the idea that the embodiment design of a product is to be combined using various elements or that an existing structure is to be analysed by defining active surfaces and pairs of active surfaces. Active surfaces and pairs of active surfaces are to be understood in this context as the basis for the functionality of a technical product. If a local change in effect is taken into account, a working space is created. Roth also speaks in this context of structural function elements, which he then equips with design elements. This approach is also referred to in a more modern way as *Generative Design, which* refers to and is particularly used in computer-aided tools, augmented by the technologies of *Features* and *Constraints.*

Systematic Variation

If the embodiment design of a technical system is to be determined during the design process, the existing solution of another product can also be used as a reference. By applying systematic variation, a new design space can be derived based on the existing solution, with which an adapted and suitable embodiment design for the new product can be developed. The only limitation in this approach is the available development time. While this approach can always develop a new solution, it cannot guarantee that the new solution will achieve an improvement.

The following illustration shows examples of various structural elements and possible variations of the design features, number, shape, topology, and dimensions (Fig. 10.3).

When designing technical products and especially in the context of embodiment design, compromises often have to be made. These are frequently necessary as an increasing number of requirements also increases the probability and frequency of conflicts of objectives when determining optimal features. To both discover and resolve such conflicts of objectives, a methodical approach to the optimisation and calculation of technical components is essential. The weighting of the respective objectives can usually only be done iteratively. To be able to make calculations, quantitative assumptions about geometries and materials of individual components must be defined in the design phase. Even during the creation of rough drafts, model-like calculations can help determine optimisations or main dimensions. With the help of such methods and calculations, various objectives listed below can be pursued and achieved:

- Determine main dimensions
- Carry out conceptual optimisation and simulation
- Document proof calculation.

The following illustrates the possibilities of numerical optimisation in combination with additive manufacturing using an example. Further aspects of lightweight construction are also discussed.

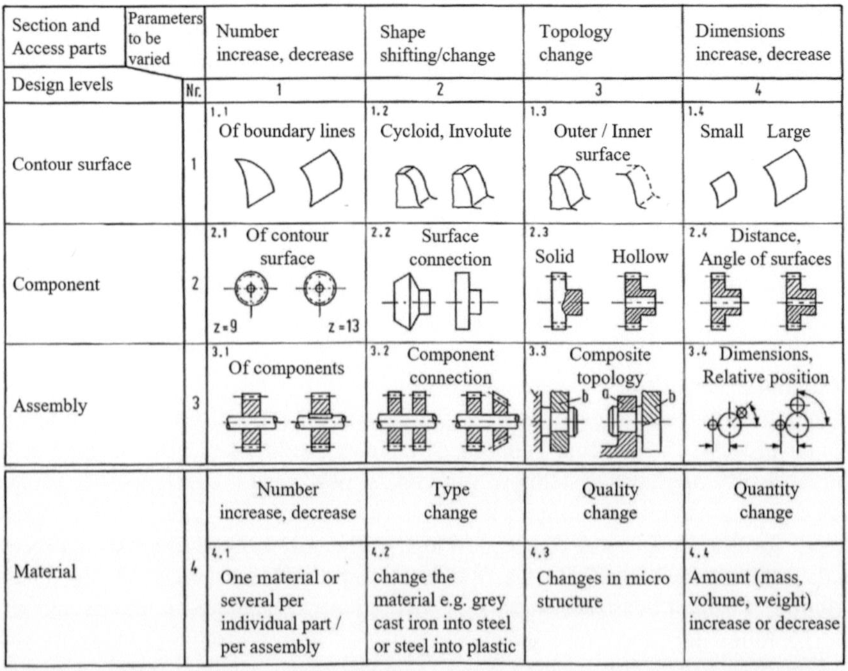

Fig. 10.3 Shape variation operations according to Roth [35]

Topology Optimisation

Topology optimisation is a special iterative numerical method from the context of structural optimisation. The procedure is fundamentally described by determining an optimal material distribution for a given installation space. The decisive design variable is a pseudo-density of the material, whereby, in an additional step, the material areas with very low density are removed, and the part is thus defined in its shape and topology. As part of this mathematical optimisation, an optimal shape of a component to be designed can be calculated for given boundary conditions. The method addresses the optimal topology and shape finding as well as the design objectives of direct and short force transmission described in Chap. 9. Figure 10.4 shows the process of topology optimisation in the context of additive manufacturing using the example of a bicycle crank.

The optimisation and redesign initially lead to weight optimisation with almost the same rigidity by maintaining the outer contour but integrating internal cavities. In the second step, the part becomes larger, thus adapting to the maximum available installation space. As a result, the bending stiffness is additionally increased due to the shape and dimensions, and further material can be removed. There is a weight-related factor of 3–1 between the first and third versions. By optimising material distribution and adding internal structures, the component weight is significantly reduced. As a result, energy can be saved during the lifecycle, as less

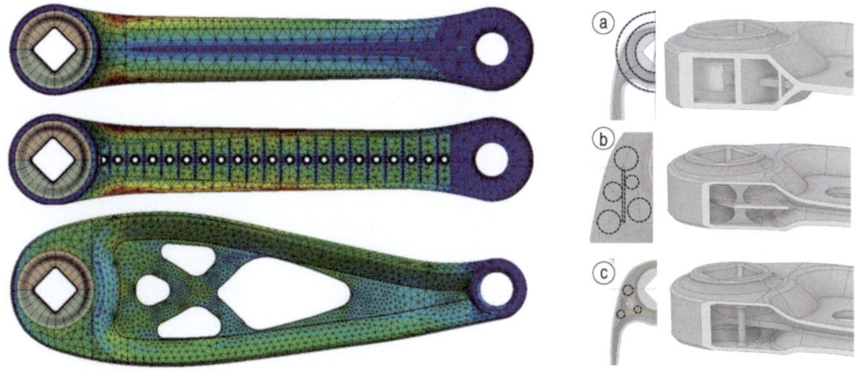

Fig. 10.4 Topology-optimised bicycle crank: on the left, various iteration stages from top to bottom; on the right, corresponding internal structures [101]

mass needs to be accelerated and moved during operation. The reduced use of material also means that less raw material is needed, and the manufacturing process requires less energy. In summary, topology optimisation is an effective and efficient approach to saving material in the embodiment design of components and reducing energy consumption during the lifecycle. Factors of sustainability can thus be specifically included in the design process and lead to an improved and more efficient result compared to conventional methods.

Lightweight Design

Material savings can be enabled through various working methods. A widely used method is lightweight design, which includes approaches such as material substitution, optimisation of material distribution, the use of internal structures, and the use of composite materials. The targeted and thus more economical use of material is becoming increasingly important due to dwindling raw material sources. By analysing the process chain, unnecessary material consumption can be identified and prevented. This reduces resource consumption and often also the subsequent costs for fuel, other operating materials, and disposal. Many other product properties, such as security and speed, dynamic properties, or the payload, are significantly determined by the weight. An example of this can be seen in aircraft construction, where the vision is to construct a complete product from lightweight materials and optimised structures. The aim is to enable maximum payload with minimal empty weight, in order to reduce energy consumption and thereby both costs and environmental impact, while simultaneously increasing the range of the aircraft. For today's large aircraft, the take-off weight for a transatlantic flight, for example, is composed of approximately one-third each of aircraft, kerosene, and payload.

Adhering to the following guidelines for embodiment design for lightweight construction leads to efficient and weight-appropriate designs:

- As direct as possible force transmission and force balance;
- Realisation of the largest possible area moment of inertia or resistance moment;
- Fine structuring of structures;
- Utilisation of the natural support effect through curvature;
- Targeted stiffening of constructions in the main dimension directions;
- Preference for the integrative principle;
- Incorporation of cavities;
- Absolute exploitation of the construction (uniform stress).

Motor vehicles with reduced own weight require less energy, especially when accelerating. In addition, less fine dust is released when braking, as the brakes are less stressed. The same applies to engine technology. Although energy can primarily be saved through improved combustion technology or more efficient control of the electromagnetic field, many engine components also have to be accelerated and decelerated. Therefore, pistons and connecting rods, for example, should be designed as light as possible.

In a fully automatic washing machine, the power consumption is based on the drive of the rotating parts and the heating of the washing water. This results in various approaches to saving, such as lowering the washing temperature, reducing movements, using lighter components, and improving insulation. This poses challenging tasks if the main functions of washing and spinning are to be maintained.

Further guidelines for embodiment design can be found in the standard work *Pahl/Beitz* [102].

10.2 Additive Manufacturing

Additive manufacturing refers to the automated production of components based on digital models, by applying material voxel- or layer-wise and connecting them according to the digitally defined component geometry. The material connection is made by welding, melting, gluing, UV curing, or similar methods. Life cycle assessments of additively manufactured components show the potential of this technology throughout the product life cycle with regard to sustainability aspects. This applies equally to the realisation of effects, production, logistics, as well as energy efficiency and resource conservation during the usage phase, maintenance and repair, and possibly recycling. Both the development of processes and machines for additive manufacturing and the development of materials, as well as the corresponding fields of application and markets, are still subject to a dynamic development process. This leads to constantly new solutions and fields of application, but also cost reductions and reliability gains.

The industrial application area of additive manufacturing can currently be divided into three approximately equal sectors: prototype construction, fixture and tool construction, and direct manufacturing. The process is primarily used in the production of prototypes or products in small quantities. Materials for additively

Outline section			Main section		Access section							
State of matter	Shape	Bonding mechanism	Designation/label		Poylmer	Metal	Caramic	Layer thickness [µm]	Support structure	Req. build chamber	Multi material capacity	Build volume
Solid	Powder	Binding	Powder bed fusion	Laser sintering	X	X	X	< 10		Yes	Limited	medium
				Laser Powder Bed Fusion		X		10-100	X	Yes	Limited	medium
				Electron beam melting		X		10-100	X	Yes	Limited	small
			Directed energy deposition	Laser metal deposition (LMD)		X	X	>200		No	Yes	medium
		Binder	Binder jetting	3D-Printing / Binder Jetting	X	X	X	> 100		Yes	No	large
	Filament	Melting	Material extrusion (MEX)	Fused deposition modeling	X			10-100	X	No	Yes	large
				Fused Layer Modeling								
				Laser wire deposition		X		>200		No	Yes	medium
	Foil	Gluing	Layer lamination	Layer Laminated Manufacturing	X	X	X	10-100	X	Yes	No	large
Liquid	Liquid	UV	Vat photo polymerization	Stereolithography	X		X	< 10	X	Yes	No	large
				Two-photon polymerization	X		(X)	< 1		Yes	No	small
				Digital Light Processing	X		X	10-100	X	Yes	No	medium
			Free beam material deposition	Multi-Jet Modeling	X		X	10-100	X	Yes	Yes	medium

Fig. 10.5 Design catalogue of additive manufacturing processes

manufactured components mainly include metals and plastics, but also ceramics or glass. Recent research also focuses on material gradations and multi-material components. Figure 10.5 shows an overview of relevant additive manufacturing processes, associated characteristics, and process-specific features, presented in a design catalogue.

Applications of additive manufacturing can be found in the production of individual implants such as dental prostheses, ear inserts, hearing aids, or individualised exoprostheses. Further selected examples are listed below:

- Repairs of high-quality capital goods by contour-following welding are state of the art, for example, for the repair of turbine blades.
- Additively manufactured tool inserts with optimised internal cooling structures are frequently used to reduce internal stresses and distortion of components or to achieve shorter cycle times.
- Large gearbox housings can be produced directly from metal using additive manufacturing without conventional tools and complex moulds.
- Highly integrated lightweight components for aerospace, such as valve blocks, are used for various space missions.
- Structures made from lunar rock could be manufactured directly on the moon.

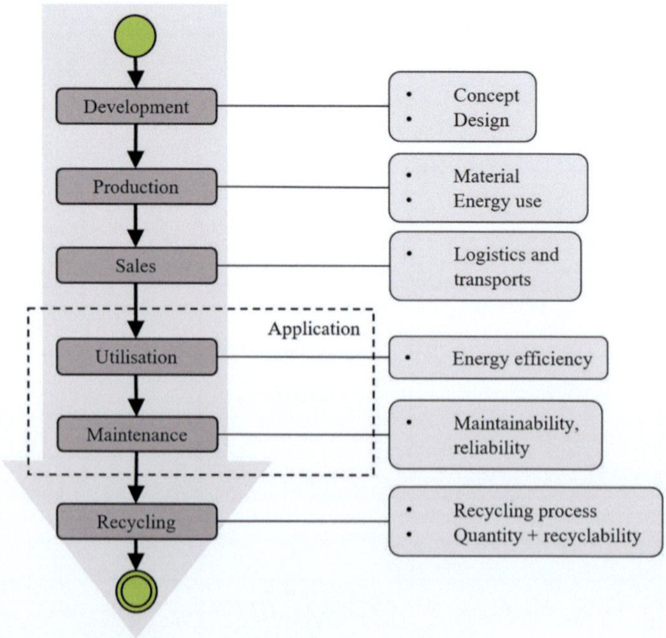

Fig. 10.6 Sustainability potentials of additive manufacturing

Figure 10.6 assigns the sustainability potentials of additive manufacturing to various life cycle phases of a product.

- **Development**: The optimised realisation of effects through design freedoms allows, for example, to design according to force flow, optimise component topology, minimise losses, integrate internal structures and flow channels, better control vibrations, optimise heat transfers, achieve high functional integration, and much more. With the advancement of technology, it will be possible in the future to optimise component properties even more specifically using multi-materials or graded materials.

- Tool-less manufacturing and the implementation of near-shape geometries, thin walls, and internal cavities can save both time and material in the **manufacturing process**.
- In **logistics** and **transport**, there are numerous potentials. Just-in-time manufacturing on-site can reduce time and energy expenditure for transport and storage. Additionally, handling technology and packaging can be printed individually, simply, and quickly.

- Energy savings in **usage** can be based on a lightweight design, but also on better realised effects. Further potentials arise from design, ergonomics, and mass customisation in general.
- What applies to logistics in general also applies particularly to **servicing** (maintenance and repair). Additively manufactured spare parts can be produced just in time and directly on site, significantly reducing the time and energy required for transport and storage.
- In the area of recycling, especially of process waste, there are extensive research activities. In general, it can be stated that material can usually be reprocessed and reused in a subsequent production cycle.

Figure 10.7 shows design goals and additively manufactured components from exemplary projects.

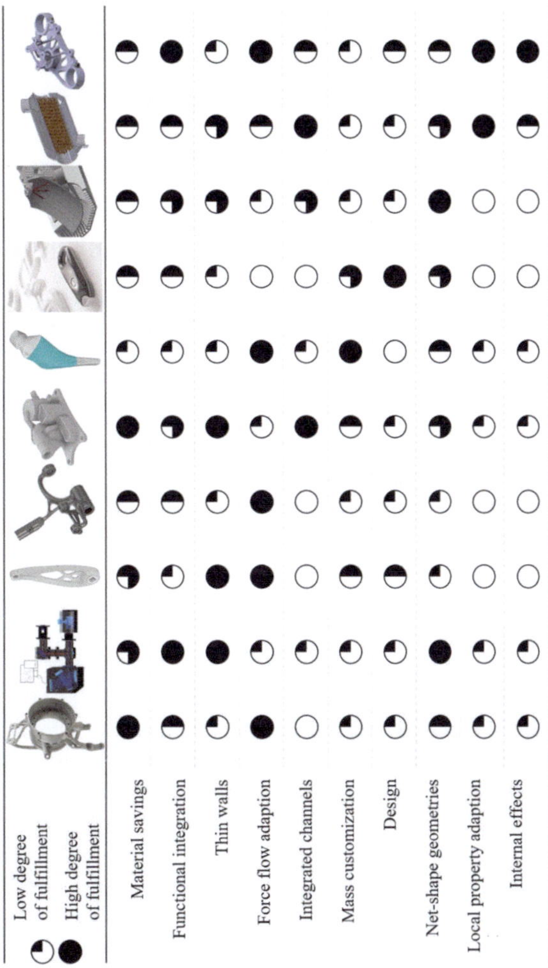

Fig. 10.7 Design goals and component realisation [101]

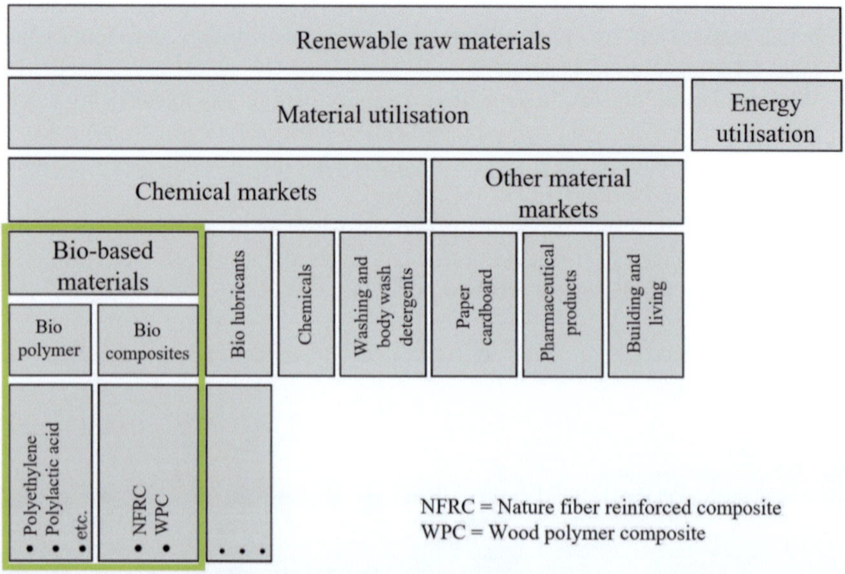

Fig. 10.8 Usage of renewable raw materials; own representation based on Saulich [104]

10.3 Bio-based Materials, Composites and Wood

As a general rule, a material can be described as bio-based if it is a product derived from biomass [103]. Biomass is a substance of biological origin, without geological or fossil sources, such as animals or plants. Renewable raw materials are biomasses that can be used both materially and energetically in various forms, excluding food. A bio-based product is thus created by the material usage of these renewable raw materials and can consist entirely or partially of biomass [104]. Fig. 10.8 provides an overview of the usage of renewable raw materials.

The focus will be on bio-based composites and, in this context, explicitly on bioplastics. These are marked in green in Fig. 10.8.

Composites

Composites are always referred to when at least two materials are combined into one material. This usually consists of a plastic matrix and another material component, such as natural reinforcement fibres or wood. The plastic matrix in today's applications usually consists of thermoplastic or thermosetting materials. As soon as biological components are integrated into the composite, it can be referred to as a bio-composite. This context is schematically represented in the following Fig. 10.9 and expanded by typical variants in the production of composites.

First, natural fibre reinforced plastics (short: NFRP) are considered. These material composites consist of naturally occurring reinforcement fibres embedded in a polymer matrix. The concept is already being successfully used to improve the

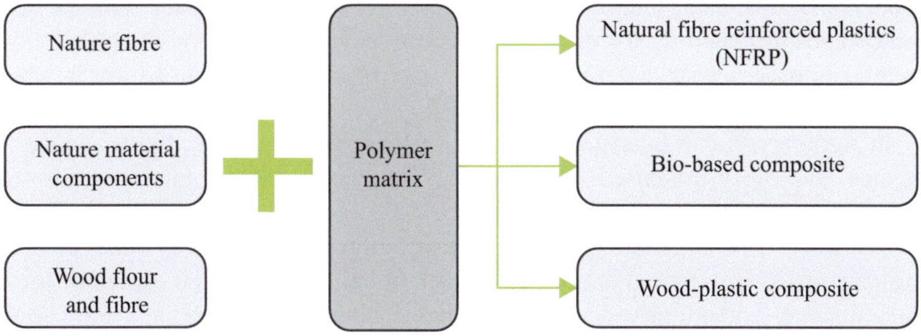

Fig. 10.9 Structure and variants of bio-composites; own representation based on Saulich [104]

stiffness, strength and impact toughness of materials. However, such composites pose a number of challenges compared to traditional alternatives due to the differences between natural fibres and technical fibres. These differences are particularly evident in geometry, chemical structure, quality variations of natural fibres, reliability under environmental conditions, and lower strength compared to glass or carbon fibres. This fact must be taken into account separately in the design and shaping process [105]. For natural fibre-reinforced plastics, fibres such as hemp, flax, jute, sisal and others are used. These materials provide stability and rigidity with low mass and medium mechanical load capacity. Natural fibres are renewable raw materials and offer advantages such as high availability, low energy requirements in production, and uncomplicated disposal when using a suitable bio-based matrix in their composites. These include, for example, starch, sugar, cellulose or plastics such as polyethylene [106]. Injection moulding is a cost-effective manufacturing process suitable for up to a maximum fibre content. For even higher strength requirements, processes used for other fibre composite parts are also used—laying, winding, and curing of resins. Examples of applications can be found in the field of lining and insulation technology in automotive construction.

Another focus will be on so-called Wood-Plastic-Composites (short: WPC). These materials, also known as wood-polymer materials, are based on wood flour and short fibres, which are bound by bio-based plastics and are thus suitable for the production of inexpensive components of low strength that can be manufactured by injection moulding or extrusion. Exemplary products made of WPC are terrace floors, floorboards or insulations [107].

Other examples of bio-based materials are plant oils such as rapeseed oil, sunflower oil, palm oil or castor oil, which can serve as fuel or the basis for biological lubricants. However, with suitable catalytic reactions, they are potentially also used for the production of, for example, ester groups for the production of polyethylene (PE).

Research projects aim to develop Bio-PPT and Bio-PBT with cellulose fibre reinforcement for technical use, which can exceed the properties of fossil and glass fibre-based materials. Of particular interest is the field of lightweight construction, which is particularly important in the automotive and electronics industries to save energy and facilitate handling. In these research projects, product samples and bio-composite materials for different application areas with comparatively good technical properties were developed and tested. However, widespread use in large quantities has not yet been confirmed [108].

In summary, it can be stated that in the field of bio-materials, bio-based composites and other bio-based raw materials, there is considerable potential for sustainable or more sustainable embodiment design of products. The examples chosen here show that the technical usage of bio-based materials is quite feasible in many areas. However, there is often a lack of validated material characteristics, guidelines for embodiment design and dimensioning rules, as well as clarification of approval issues and reliability aspects. There is also only limited experience available regarding long-term environmental impacts.

Wood

Historically, wooden constructions have played a significant role in technical development. Windmills, ships, aeroplanes, and vehicle bodies were made of wood for centuries and still are to some extent. While wood continues to be a significant raw material in construction, the question arises as to the potential of this material in mechanical engineering and how components should be designed accordingly. The use of wood as a material can address problems such as the scarcity of fossil raw materials or global warming. The continuous cycle of planting, growing, felling and utilising wood can be a crucial part of circular, sustainable development processes. Through various approaches and efforts to utilise the advantages of wood as a construction material, mineral resources can be substituted in the long term, making a significant contribution to sustainability [109].

At first glance at sustainability-relevant aspects of modern wood materials, various advantages over conventional alternatives such as metals or plastics can be noted. Wood-based materials, such as modern wood materials or wood-polymer composites, are technically stable, resilient and characterised by inherent vibration damping and reduced noise emission at a comparably low weight. Depending on the application, a reduction in component weight of up to 50% compared to metallic materials can be achieved. This results in potential for lightweight design applications. Further advantages result from the corrosion resistance and comparatively low thermal conductivity and thermal expansion of wood. The BMWK estimates in a study that greenhouse gas emissions can be reduced by 90% or more when using and manufacturing wood materials compared to conventional solutions. Furthermore, it is described that the consumption of fossil raw materials can be reduced by 40–59% when bio-based materials like wood are used. It is also noted that resource efficiency can be significantly increased by avoiding waste and utilising wood-based waste [110].

The need for wood to become more relevant as a modern material and consequently be more intensively researched is evident when looking at the current state of research and science. As part of the joint project "Wood-based materials in mechanical engineering (HoMaba)", the use of wood materials in mechanical engineering was intensively investigated and scientifically processed by nine research institutes. The main concern of the project was to research wood as a sustainable and available construction material more intensively, in order to contribute to a more climate-positive industry and to explore opportunities for sales markets in plant and mechanical engineering. As part of various sub-projects, characteristic values, approaches and safety concepts were adapted to the requirements of mechanical engineering in such a way that reliable statements about components made of wood materials could be derived. The specific contents of the sub-projects were the determination of characteristic values for wood materials, the development of calculation and simulation concepts, the development of testing methods, the validation of the results using suitable demonstrators, the creation of characteristic value databases and knowledge transfer [111].

Despite the potential of wood as a building material, its application today is hesitant and often not successful. For example, in 2012, a wooden wind turbine tower named *Timber Tower*, the first of its kind, was erected by *Vensys Energy AG* in Hanover. The prototype, recognisable by its faceted cross-section, stands directly on the A2 motorway. However, due to the significant increase in performance and size growth of modern wind turbines, all new plants are built with towers made of steel or concrete-steel combinations for load-bearing reasons, to ensure the required stability. Nevertheless, there are various efforts, mainly due to ecological advantages, to use the concept of wooden towers for further wind turbines [112].

Another example comes from the logistics sector. The company *ligenium* processes wood as the main material and produces load carriers in this context. The focus is on the use of wood as a lighter and more environmentally friendly material, enabling CO_2-neutral production. The company uses wood to produce transport carts for moving various components in the field of intralogistics. As a result, the cart only weighs 145 kg, while comparable metal models weigh around 400 kg. This weight saving not only leads to lower transport costs, but also to a reduction in energy consumption and CO_2 emissions [113].

A prime example of the described application of wood in mechanical engineering—here in leisure sports—is the wooden bicycle concept by *Niko Alber* from 2015. In this conceptual design, known as *Timber Bicycle*, thin oak veneer forms the three-part bicycle frame. The construction uses the springy and elastic properties of the wood to absorb vibrations, achieving a comfortable suspension effect on uneven city streets. Oak is chosen for its advantageous mechanical properties, including high strength, elasticity and absorption capacity. In addition, oak is characterised by good properties in terms of compressive strength, durability and weather resistance. The availability of oak in Europe further complements the advantages of this material choice. The manufacturing method, which is based on the use of many individual oak veneer layers, allows for high design freedom and a targeted positioning of the suspension functionality in the production process. Aluminium connections are used to functionally complement the wooden frame and harmoniously integrate into the overall design. In addition to the development of this design bicycle, there are a multitude of other models, including the so-called *Street Cruisers* by the company *My Esel* or off-road optimised bicycles by *Cyclowood*.

In a concluding assessment of the material use, several factors should be critically examined. This includes the ratio of material use to achieve functionality. In addition, a comparison of the lifespan of the final product with the regeneration time of the used wood resources is relevant. The CO_2 balance also plays a crucial role in the context of sustainability. Furthermore, the possibilities of replacing wood with rapidly renewable natural materials such as bamboo, hemp and flax should be carefully examined.

In summary, it can be stated that in the context of the constructive use of wood in mechanical engineering, there is considerable potential for sustainable or more sustainable embodiment design of products. The explanations and examples provided demonstrate that using wood-based materials is both technically feasible and practical in many areas of application. This is particularly true when the emphasis is not solely on maximising performance but rather on achieving a balanced approach to all aspects of sustainability. Compared to plastic-based and metallic materials, the use of wood shows an improved CO_2 balance and an overall optimised environmental balance. Wood is also characterised by high availability and appropriate reusability. From a circular economy perspective, wood can not only be reused in a sensible way after its primary use, but it can also be biodegraded or utilised for energy.

Product Lifetime

11

The lifespan of products has a significant impact on resource consumption throughout the entire lifecycle of product generations. This results in different strategies to combat obsolescence (the deliberate shortening of product lifespan), including legal instruments. These can set requirements or incentives that enable longer use of products. Moreover, longer lifespans are possible. The determination of technical parameters offers the opportunity for better monitoring of product lifetimes. Thus, minimum requirements such as statistical lifespan specifications are firmly attributed to the products. These are checked, for example, by market surveillance authorities. Specifications of this kind are already prescribed for some product groups by EcoDesign regulations. Lighting and light bulbs are a classic example.

From the consumer's perspective, the introduction of extended manufacturing warranties would be a solution to extend the product's lifetime. Through the warranty period, manufacturers document their confidence in their own products and provide a better basis for deciding to purchase a product. Currently, the statutory warranty periods are limited to two years after receipt of the goods. For consumers, this represents a significant risk, especially for products such as large household appliances, which typically have a much longer lifespan. The current deadlines therefore, do not meet the expectations of the lifespan of these products. Better conditions should be created to promote repairs by commercial enterprises and private initiatives. There is a need to develop proposals for improving legal requirements and optimising cost structures. The focus should be particularly on the availability of spare parts, including their delivery times, pricing and access to relevant repair and maintenance information, as there is often a need for improvement here. To promote durable products and to make planned obsolescence unattractive for manufacturers, the deliberate shortening of product lifespan was introduced as a criminal offence in France. There, companies can be legally prosecuted if the lifespan of a product is deliberately shortened to achieve faster new purchases. If

R. Lachmayer et al., *Methodology for the Development of Sustainable Products*,
https://doi.org/10.1007/978-3-662-72442-2_11

a punishment for "planned obsolescence" were introduced in Germany, this would primarily have a symbolic meaning.

However, it remains of crucial importance that particularly long-lasting products are safe in terms of potential hazards and damages to people, facilities and the environment and do not pose a threat. The following defines some terms in the context of reliability that are relevant for this chapter.

- **Threat**: These are potential events that can have negative effects if they occur.
- **Damage**: Damage includes impairment to human health, the environment, and material and immaterial losses.
- **Risk**: A risk is the combination of the probability and consequence of an event.
- **Reliability**: Reliability describes the combination of fulfilled properties and performances that ensure a product can always be used reliably and as intended over its planned or expected lifespan.

At the beginning of a development, appropriate **reliability**, **availability**, and **safety objectives** are always set. Criteria for this are acceptable risks, tolerable failure frequencies, downtimes, and the associated lifecycle costs.

When basic technologies become obsolete and are replaced by new, more efficient technologies, or when personnel no longer have the necessary knowledge for the safe operation of plants, additional challenges arise. Technology cycles can vary considerably, with some, such as in the semiconductor industry, occurring within a few years, while others in the nuclear industry can last several decades. This raises the question of how to deal with technologically outdated product generations and when they must be decommissioned at the latest.

From an economic perspective, the question of the optimal product lifespan arises, particularly in evaluating current costs against potential long-term benefits. Determining this optimal lifespan and its possible extension, taking into account ecological aspects, is a complex challenge in the face of continuous technical and human progress. Figure 11.1 explains the terms' reliability, security, availability, and maintainability in this context.

The following will examine the aspects of reliability, servicing, continuous improvement, and the vision of the "eternal product" from a technical perspective.

11.1 Reliability

The reliability of a technical product describes how reliably a function assigned to the product or a system is performed within a certain period of time. As it is subject to a stochastic process, reliability can be described qualitatively or quantitatively by means of survival probability, but it is not directly measurable and is not identical for identical products in reality.

Reliability is a fundamental characteristic of all technical products, so no technical product is completely safe from failures and will ultimately fail at some

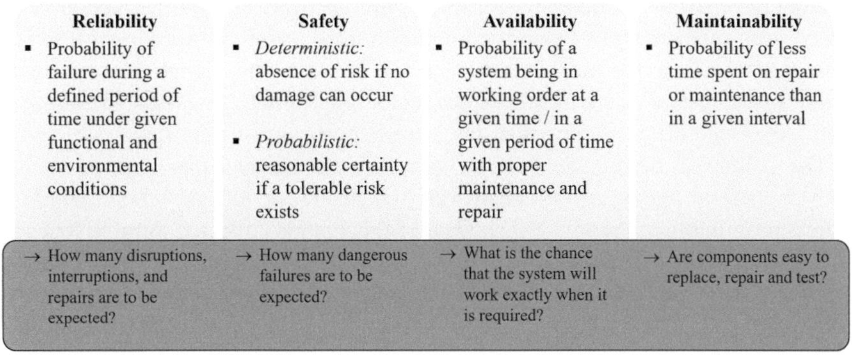

Fig. 11.1 Product behaviour over the lifespan

point. The reliability of a product can be determined or estimated either computationally, by determining the frequency of failure, or analytically, by deriving the reliability values of the individual components of the product. For simple technical devices, the empirical approach is usually chosen. More complex systems, however, (automobile, power plant, railway, aircraft) can only be considered analytically in terms of dangerous conditions in a reliability proof. For the reliability analysis, models such as the fault tree and event tree model, as well as the failure mode and effects analysis (short: FMEA), are used. These models represent the failure structure of an overall system and enable their calculation. The reliability or failure probability of the overall system is calculated based on the failure rates of the individual components. Complex reliability analyses require an experienced team, careful planning of all necessary steps, a suitable reliability database, and special software. The procedure and the details are comprehensively described in the VDI 4003 and IEC 60,300 standards. For example, VDI 4003 also provides a comprehensive overview of the analytical methods used today.

Software reliability refers to the probability that a software application will function without errors over a specified period of time under specific environmental conditions. Unlike hardware, software does not wear out because it is immaterial. Therefore, the error rate of software is not dependent on the age of the software or the frequency of its usage. Basically, three different types of software errors are distinguished:

- **Faulty requirement**: This is where errors occur in the software requirement.
- **Design error**: This is a faulty design in relation to the specified requirement.
- **Programming error**: Errors in programming that occur in relation to compliance with the software conceptual design.

Before testing software, it must always be implemented in hardware. When identifying errors, it is therefore necessary to precisely determine the cause to determine whether the error is caused by the hardware, the software, or their interaction. Software errors that were not discovered and corrected during development tests are present as hidden error mechanisms and only occur under certain system conditions. As the system usage increases, the probability of detecting these latent errors increases. Once they are corrected, the error rate of the software decreases.

The collection of reliable data on various wear mechanisms of components, such as thermal ageing, tribological wear, corrosion, or loss of time strength, is crucial for reliability analysis [102]. Data of this kind is collected through material tests and extrapolated or collected from operational experience. In this process, the frequencies and causes of failures of comparable products are systematically analysed. Today, synthetic data also plays an increasing role in the validation of material cycles, especially of complex systems.

Data collections show that technical products typically go through three different failure phases during their lifetime. Initially, so-called early failures occur, which can be attributed to initial weaknesses in the design and can be corrected with increasing operational experience. This is followed by the usability phase, in which the failure behaviour remains low and largely constant. At the end of the products' life, wear failures accumulate, leading to an increased failure rate up to complete failure with 100%. This typical behaviour, often represented in the form of a characteristic distribution curve, known as the *bathtub curve*, is visible in the following Fig. 11.2. The Weibull distribution can be used to model the reliability distribution, which is characterised by decreasing, constant, or increasing failure rates, as its flexible parameterisation can represent these different phases.

The Mean Time Between Failures (short: MTBF) is a measure of the reliability of repairable units such as assemblies, devices, or systems. If the failure rate is constant and the reliability distribution is exponential, indicating the presence of random failures, the MTBF value can be calculated by the reciprocal of the failure rate. This principle also applies to the key figure of the Mean Time To Failure (short: MTTF), which, however, is used for non-repairable units.

One method for determining lifetimes or error rates is the accelerated ageing procedure. This is done according to the Arrhenius or the Eyring method, which is

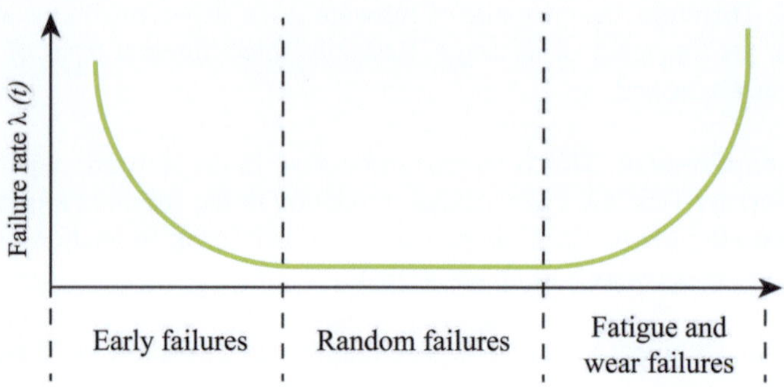

Fig. 11.2 Characteristic curve of the reliability distribution

often used by component manufacturers, particularly for small component populations. In comparison, the Highly Accelerated Life Tests (HALT) and End of Life Tests (EOL) are noteworthy. The method is defined in various standards—some of these are listed below:

- ISO-Standard 18,921:2008: "Image Recording Materials—Compact Disc (CD-ROM)—Method for Determining Life Expectancy Considering the Effects of Temperature and Relative Humidity";
- Standard ECMA-379 (identical to ISO/IEC 10,995:2008), "Test Method for the Estimation of the Archival Lifetime of Optical Media";
- USA—National Institute of Standards and Technology
- (NIST): "Optical Media Longevity Study".

In information technology, determining the lifespan of digital storage media, such as hard drives, USB sticks, CDs, DVDs, magnetic tapes, and floppy disks, is becoming increasingly important for the long-term archiving of digital information. As these storage media are based on very different technologies, they exhibit different failure mechanisms and thus varying lifespans.

The type and frequency of failures, as well as limitations of availability, can be influenced and largely planned. The following measures can be used to increase the reliability and extend the service life of products and to avoid the consequences of failure:

- Systems architecture and operational management;
- Increasing safety margins in design;
- Control of stresses (avoidance of misuse);
- Use of higher quality materials and coatings;
- Quality management;
- Use of proven and qualified components;
- Use of redundant and diverse components;
- Measures for self-detection and correction of errors;
- Application of the fail-safe principle;
- Testability of components and system complexes;
- Preventive qualified maintenance/servicing;
- Design concerning ergonomics, user-friendliness, and error-tolerant handling;
- Data management and evaluation of feedback;
- Monitoring, disturbance, and emergency management.

In practice, with complex products or systems, such as motor vehicles, typically several of the previously mentioned wear mechanisms are effective. When classifying, it must be coordinated how the various wear processes acting over time can be coordinated, depending on the number of repairs or maintenance needs, so that a product reaches a uniform end of life in all its components. Alternatively, the series should be designed so that no additional disposal problems or errors arise from a sustainability perspective.

After these introductory notes on reliability, the following will deal with possibilities for concrete reliability analysis. The FMEA method will be presented and described as an exemplary procedure.

The **FMEA** is a method to identify possible failure modes, their causes, and their effects on system behaviour and to optimise the security of systems already during development. The analysis can begin as soon as the system is defined to the extent that it can be represented as a function structure or architecture. The basic idea is to identify all conceivable failure modes for any system, subsystem, or component, and to show the possible failure consequences and causes. The FMEA is a risk analysis integrated into the development and process planning of new products. It is an important part of quality assurance before series start-up and the realisation of safe products in the field. Figure 11.3 shows the procedure for developing an FMEA and the quantitative assessment of the solution based on a risk priority number.

After describing the system or operation and its function, potential failure modes are identified, and measures for error detection and the probability of detection (E) are estimated. Subsequently, the causes of failure are determined, and the probability of occurrence (A) of the error and measures for error prevention are defined. Finally, an estimate of the effects of the error is made, including measures to limit the effects and the assessment of the severity of the effects (B). The occurrence probability, detection probability, and severity of the impact can

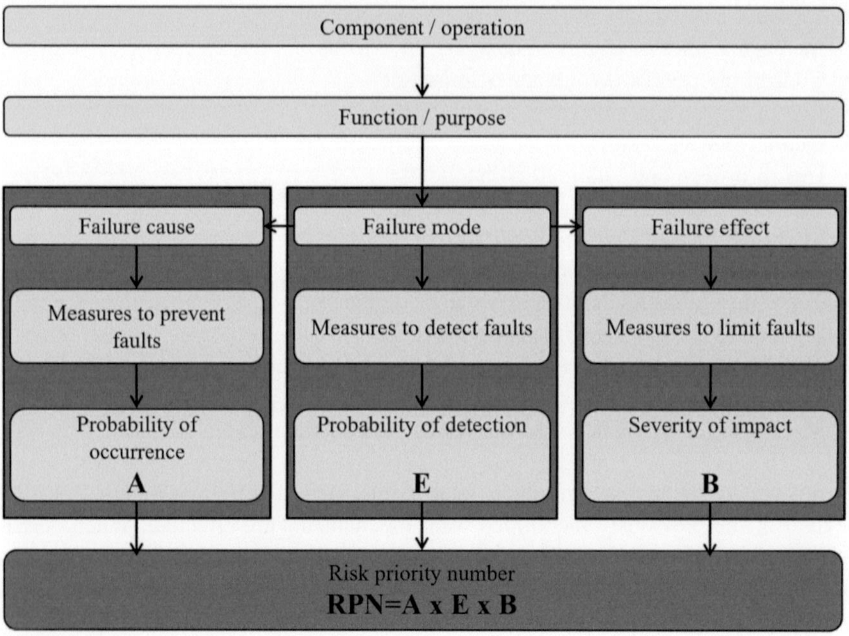

Fig. 11.3 Failure mode and effects analysis (FMEA)

be quantified for each possible error using a risk priority number, which is formed from the product of A, E, and B. The assessment for A estimates how likely the occurrence of the failure cause is. It addresses the question of whether it is a rather hypothetical error or whether it has already occurred frequently in practice. The occurrence probability is weighted as follows:

- unlikely = 1
- very low = 2–3
- low = 4–6
- moderate = 7–8
- high = 9–10

The assessment score B also describes the significance of the failure sequence. The endangerment of individuals leads to a high assessment, while, for example, minor comfort impairments receive a correspondingly low score. The assessments of environmental damage are comparatively rarely discussed. The significance of the impact is weighted as follows:

- barely noticeable impact = 1
- insignificant errors (minor strain on the customer) = 2–3
- moderately severe error = 4–6
- severe error = 7–8
- extremely serious error = 9–10

With the detection score E, it is determined how certain the discovery of the failure cause is before delivery to the user. The detection probability is assessed as follows:

- high = 1
- moderate = 2–5
- low = 6–8
- very low = 9
- unlikely = 10

From a methodological perspective, the goal of every development should be to achieve the lowest possible risk priority number. For this purpose, extensive constructive measures are available that support optimisation through error prevention, error detection, and limitation of the effects.

The theoretical description is illustrated in the following Table 11.1 using an example of an FMEA carried out on the well-known example of the kettle system. The exemplary contents and the values entered for the corresponding categories do not claim to be complete, but merely serve to clarify the theoretical principles described.

Table 11.1 Exemplary FMEA for the kettle system

Component	Possible failure	Cause	Impact	A	E	B	RPN	Measures to reduce risk
Heating element	The heating element overheats	Thermostat defect or lack of water contact	Fire hazard, the device gets damaged	3	5	9	135	Install a high-temperature fuse
Thermostat	The thermostat does not switch off	Mechanical defect or calcification	The heating element overheats	4	3	7	84	Suggest regular maintenance and descaling
Lid	The lid does not close correctly	Deformation or contamination	Leakage of hot water/ Steam	2	4	6	48	Use robust materials
Water level indicator	The water level is displayed incorrectly	Condensation or covering in the viewing window	User misoperation	3	6	4	72	Optimise display (e.g., illuminated, internal)
Power cable	The cable breaks or is damaged	Repeated bending or material fatigue	Short circuit or electric shock	3	3	8	72	Use flexible and robust cables
Housing	Housing becomes hot	Insufficient insulation / Material defect	Risk of burns	3	4	8	96	Use improved insulation materials
…	…	…	…	…	…	…	…	…

11.2 Maintenance/Servicing

Typically, a product or its associated parts, as well as operating materials, lose performance over time. Figure 11.4 illustrates how the performance of a product changes over time due to usage. After a certain period of use, the product falls below the minimum requirement for reliability and is therefore no longer reliable. If performance continues to decline from this point, the minimum requirement for safety is not met, and the product is no longer classified as safe. Various measures can be used to increase or restore performance. The figure shows repair and maintenance as examples. Depending on the phase of use and performance, these maintenance measures can make the product safe and reliable again. Maintenance and repairs may be necessary due to a legally regulated maintenance obligation to maintain safety, or to prevent system failures and extend the service life, or to secure the investment.

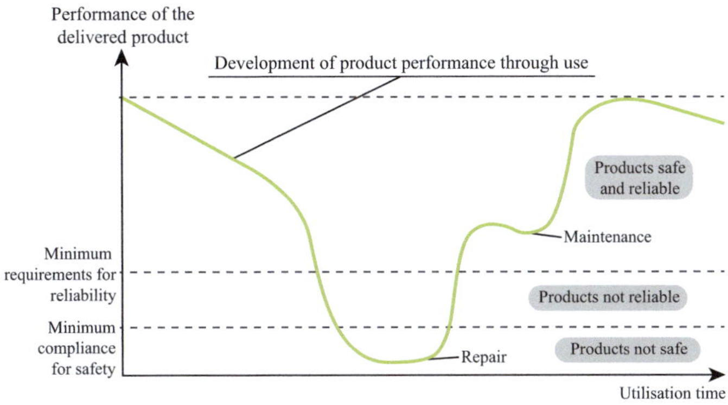

Fig. 11.4 Ensuring performance through maintenance and servicing, based on Vajna [114]

The goals of maintenance measures, such as maintenance and repair, can be:

- Increase and optimal usage of the product's lifespan;
- Maintain operational safety;
- Increase availability;
- Optimise processes;
- Reduce disruptions.

To achieve these goals, regular inspections, maintenance, or repair tasks must be defined, communicated, and carried out. This also requires a smooth supply of spare parts and the cost-effective provision of the necessary expertise.

The most important maintenance strategies are:

- **Repair after failure (Reactive)**: No preventive servicing is carried out. The failure is accepted, and errors are corrected after they occur. A typical example is the replacement of (non-critical) lamps after a defect.
- **Planned Servicing (Proactive)**: There are preventive measures to minimise the risk of failure. These are divided into:
  - Scheduled Servicing (preventive): Regular implementation of preventive measures to proactively prevent the occurrence of faults. Exemplary measures include inspections, maintenance, or the replacement of entire components (e.g., tyre change after mileage/time);
  - Condition-oriented Servicing (predictive): Recording wear-related conditions. This can either be done permanently through sensors (for example, in the form of Condition-Based Maintenance) or irregularly, depending on the time during human inspections. Selected measures are then taken, depending on the individual's wear condition.

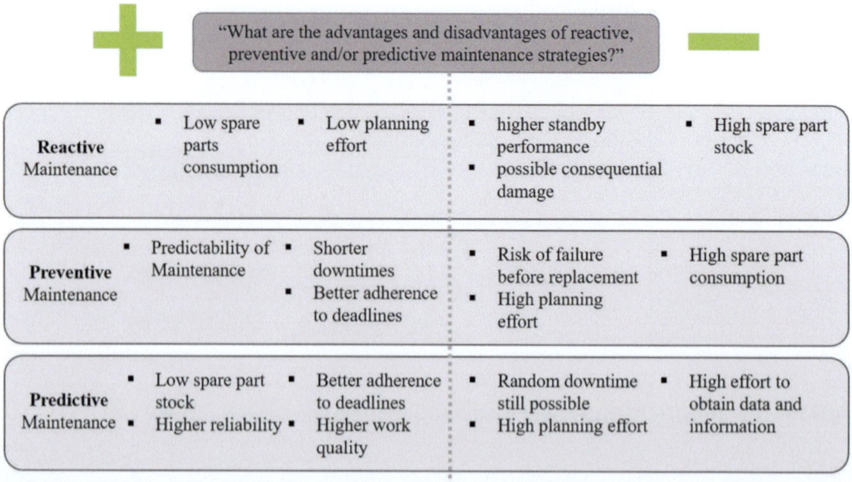

Fig. 11.5 Advantages and disadvantages of different maintenance strategies; own representation based on Biedermann and Kinz [115]

To summarise the existing relationships as well as the advantages and disadvantages of the different maintenance strategies discussed earlier, an overview and comparison of the various strategies is presented in Fig. 11.5.

For a final and concise overview of the terms used, the terms Maintenance, Inspection, Repair, and Improvement according to DIN 31,051 are clearly compiled and presented in the following Fig. 11.6.

In summary, it can be stated at this point that the selection of suitable maintenance strategies can significantly contribute to improving the environmental impacts of a product. The implemented maintenance measures can significantly reduce a product's ecological footprint, extend its lifespan, and thereby optimise the material use per performance delivered. There is also the potential to achieve improved adaptation to the specific requirements during the Product Life Cycle.

11.3 Continuous Improvement of Product Characteristics

The process of continuous improvement, also known as the continuous improvement process, is a mindset based on the Japanese philosophy of *Kaizen* that aims to strengthen the competitiveness of a company through steady improvements, which are implemented in small steps. The process can relate to the product, the process, and the service quality of a company, and is a fundamental principle of quality management. In this context, the concept of continuous improvement is also an indispensable part of ISO 9001 (requirements for quality management systems) [117]. The development of improvement suggestions by working groups is

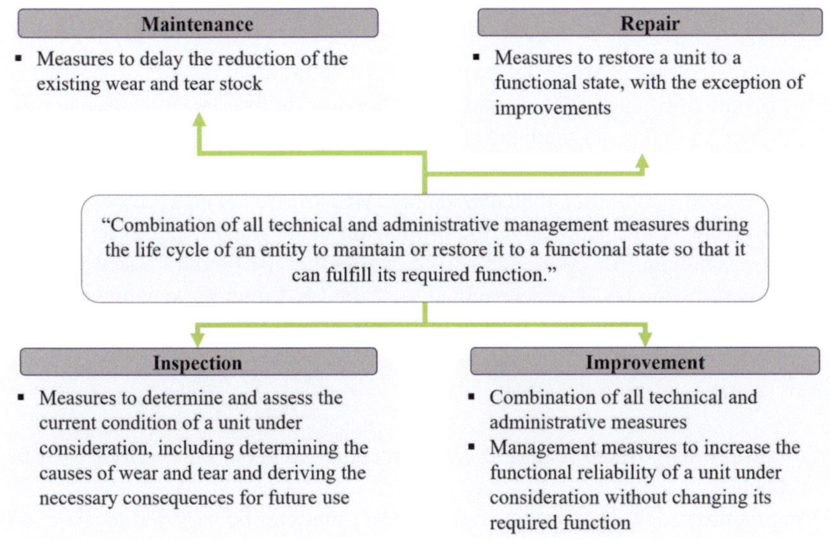

Fig. 11.6 Maintenance, Inspection, Repair, and Improvement according to DIN 31,051 [116]

usually combined with the company suggestion scheme under the term idea management. In the context of this book, the focus is on the improvement of products and their features. This involves distinguishing between the technical inheritance or the development of new product generations and the optimisation of products already in use.

The goals of continuous improvements to products and particularly complex systems in the field can be:

- Compliance and improvement of standards (laws/norms);
- Improvement of process parameters and quality;
- Reduction of consumption (material/energy);
- Optimisation of usage;
- Improvement of customer/supplier relationship.

Possible measures include:

- Software updates;
- Networking and product service models;
- Replacement of components and modules;
- Repair with higher quality materials;
- Switching to other operating materials.

11.4 From Repair to the "Eternal Product"

Within the context of R strategies (see Sect. 8.2), it has already been discussed that, from a sustainability perspective, it is advantageous to preserve the economic value of a product and reuse its modules or components, rather than merely recycling or disposing of the material.

The Material Circularity Indicator (short: MCI) in this context represents, on a scale from zero to one, how large the proportion of recycled material in a product is, with one standing for 100% and zero meaning that the product is entirely made up of primary resources. The determination of the MCI requires specific data about the material and energy flow of the products as well as their design and usage characteristics. The MCI calculation typically takes into account the following aspects:

- Proportion of recycled material: The percentage of recycled material used in a product.
- Recyclability: The proportion of the product that can be recycled at the end of its life.
- Durability: The expected lifespan of the product compared to an industry average.
- Design for circularity: Aspects of the design that promote repair, reuse, or recycling.

Table 11.2 provides some examples and their respective MCI value for illustration.

In reference to the previous explanations, even with an MCI value of zero, which indicates that a product was entirely made from primary resources, it should be noted that the energy expenditures for recycling and production must be taken into account. Therefore, it is more advantageous to restore a product's functional state through repairs and extend the lifespan of systems through reuse. The following briefly explains this relationship using the example of a washing machine.

Assuming a washing machine's lifespan of 8 years, there is a significant ecological advantage if the lifespan is extended to 16 years through three repairs. The main advantage is that the need to manufacture new devices is halved, saving resources and reducing emissions. This leads to sustainable usage of the product

Table 11.2 Exemplary everyday products and their associated material circularity indicator (MCI)

Product	Material circularity indicator
Smartphones	0.5–0.65
Furniture	0.6–0.75
Textiles	0.4–0.55
Packaging	0.7–0.85
Electronic devices	0.5–0.7
Cars	0.1–0.2

and contributes to environmental conservation. However, potential customers face various challenges when deciding on a repair. The repair costs can be substantial, and there is a risk that other parts of the washing machine may fail after the repair. Also, manufacturers or repair services usually do not extend the warranty for the entire device, increasing the risk for consumers. Thus, often repairable devices are replaced with new ones. These financial and practical uncertainties must be precisely weighed against ecological advantages.

A potential solution to keep products permanently ("eternally") usable lies in the concept of the so-called "Upgrade Factory". This implies the return of repair-needing or aged products to an industrial production, where they can be completely overhauled and updated to the latest technology (upgrade) at a low cost, to be then reintroduced to the market as new and with a warranty. The specific goals of the Upgrade Factory are:

- Extending the lifespan of products and components up to the concept of the "Eternal Product", i.e., a continuous refurbishment and reintroduction into the production process;
- Supporting the circular economy concept, particularly in terms of convincing customers by providing warranties and ensuring high-quality standards;
- The enhancement or modification of products by introducing new components and product extensions that improve the technological value of the product and maintain regulatory compliance.

The idea of the Upgrade Factory is exemplified and schematically depicted in the following Fig. 11.7 using a machine tool as an example. The majority of the mechanical components, which constitute the largest part of the installed material, are suitable for reuse without concern. In contrast, wear parts and particularly components of the machine control must be replaced due to the short technology cycles of electronics and software. The product can then either be returned to the original customer or integrated into a secondary market.

The implementation of the "Eternal Product" concept brings challenges in terms of acceptance and the development of suitable business models, as well as technical challenges. Figure 11.8 shows a compilation of these challenges for both product development and production, as formulated in the position paper of the "Wissenschaftliche Gesellschaft für Produktentwicklung" (short: WiGeP) in 2021.

As can be seen from Fig. 11.8, the challenges are formulated and declared as research needs. The overarching goal is to further develop the "Eternal Product" based on new research aspects and projects, as well as associated results, and to establish it in the long term. As part of this work, measures must also be taken to ensure broad acceptance and increase the application rate.

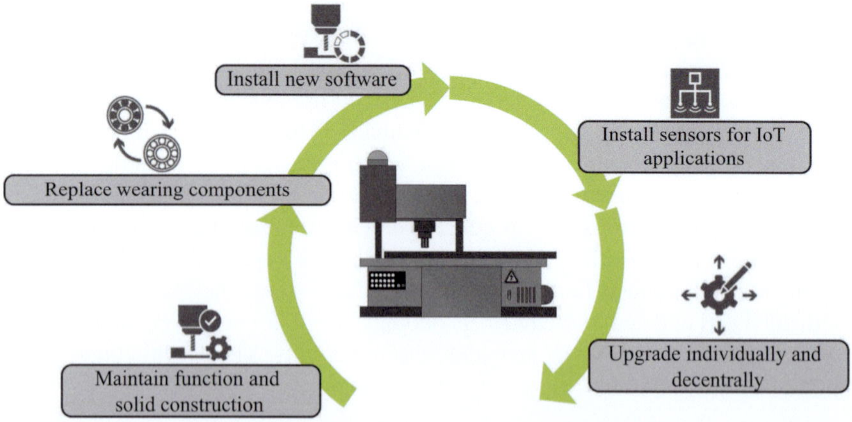

Fig. 11.7 Concept of the "Upgrade Factory" using a machine tool as an example according to WiGeP 2022 [118]

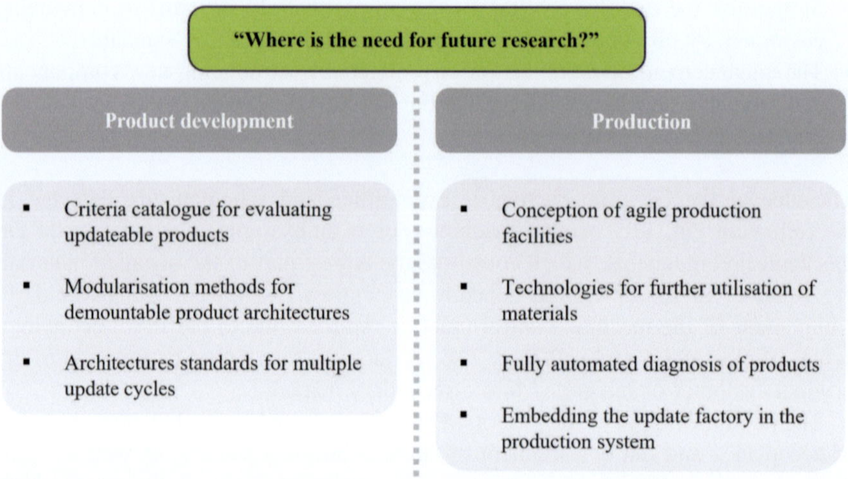

Fig. 11.8 Research needs for the "Eternal Product" according to WiGeP 2022 [118]

12 Shaping the Future with Knowledge

In this final chapter, various areas of sustainable action and doing are presented in terms of their motivating effect on product development. The idea is to substantiate the basic idea of this book, the development of sustainable products, and particularly their necessity, through various scenarios, figures, and facts. Selected statistics are used to demonstrate why the relevance of the sustainability concept—both today and in the future—must be given necessary and ever-increasing attention. The areas of sociology, ecology, and economics are highlighted in various sections, with a focus on where product developers can act, where special challenges await, and where, despite many efforts, there may be few opportunities for sustainable change. The long-term aim should be to develop sustainable products by utilizing knowledge, skills, principles, and strategies. This approach will help us address current problems with effective solutions, turn challenges into opportunities, and create a future where generations can live, work, and prosper together successfully.

Although these introductory words may give a political impression, this book does not intend to open or formulate a socio-political discussion. The statistics presented in the following serve primarily to illustrate the topics described. In this context, few or none of the aspects listed are discussed or interpreted in detail. Instead, the intention is to let the facts speak for themselves through the individual sections and figures.

Thematically, the chapter is motivated by the so-called 'Limit-of-Growth scenarios' of the *Club of Rome* from 1972 and their extension with current data. The following Fig. 12.1 shows significant trends in the areas of population, industrial output, non-renewable resources, pollution/emissions, and food.

The original report was already sensitised in the 70s that a constantly growing economy, without changing boundary conditions, quickly reaches its limits. According to Cordes and Roth [121], it becomes clear that large parts of the

R. Lachmayer et al., *Methodology for the Development of Sustainable Products*,
https://doi.org/10.1007/978-3-662-72442-2_12

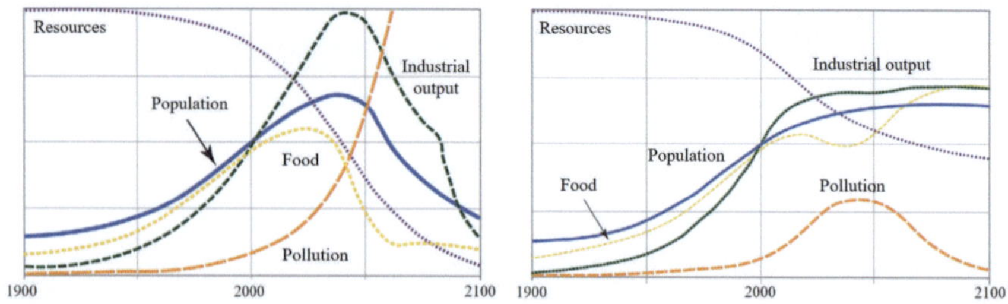

Fig. 12.1 Limit of Growth scenarios of the Club of Rome and extension with current data according to Meadows [119], figure adapted from Herrington [120]

global society are on the course of the disaster scenario shown in Figure 12.1. The graphic shows not only the negative, inevitable disaster scenario (left side) without changes, but also another course. This is a stabilised world – one in which conscious changes lead to the otherwise following negative developments being mitigated, in parts even prevented (right side). The main concern must be to use the possibilities of science and technology to change negative trend scenarios for the better. From this need, some fields of action can be derived.

As already discussed in Chap. 1, the UN, among others, deals extensively with the formulation and definition of such fields of action and recommendations in the area of sustainable development. The Sustainable Development Goals (short: SDGs) listed in Fig. 1.2 can be transferred into a hierarchical arrangement or categorisation in the context of the proposed structure. This classification is illustrated in the following Fig. 12.2 according to the graphical representation by Rockström and Sukhdev [122].

The Sustainable Development Goals are assigned to the areas of economy, social issues, and ecology. A comprehensive description of each goal can be found in various sources and will therefore not be further elaborated here.

12.1 Sociology

A first aspect in the context of sociology to be discussed here is the topic of global population. In such a discussion, the expected population growth plays a crucial role, in addition to the current population. As of spring 2025, just over eight billion people live on Earth. According to current forecasts, the world population will continue to grow steadily until 2080. This projected population development is shown in the following Fig. 12.3.

The projected increase in the world population over the next 60 years has a multitude of consequences. These include, among other things, the scarcity of food and living spaces, as well as an increase in social inequality. Furthermore, a likely increase in emissions is expected, which could contribute to a further intensification of climate change. These developments could have significant

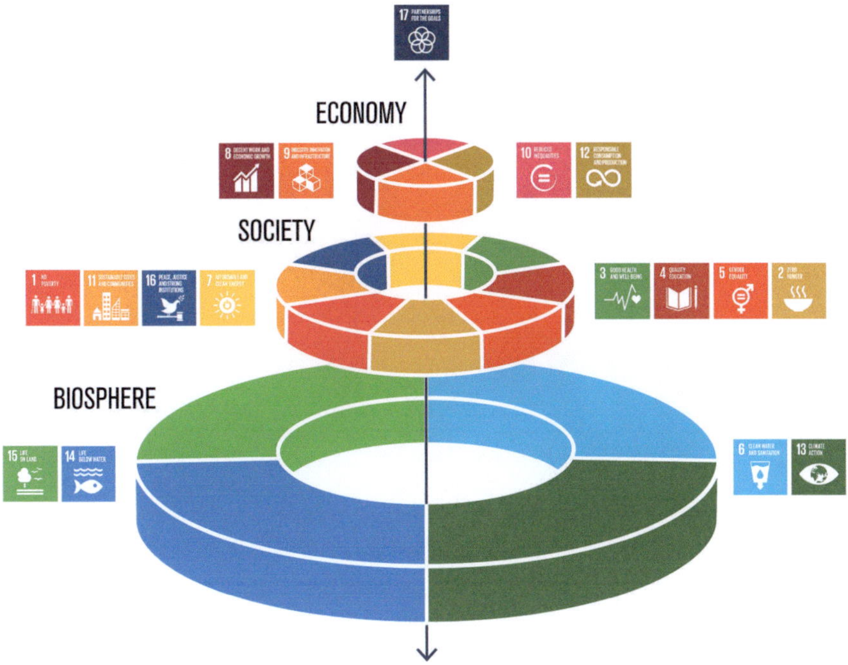

Fig. 12.2 The UN's sustainable development goals as a hierarchical representation by Johan Rockström and Pavan Sukhdev. Original: Stockholm Resilience Centre [122]

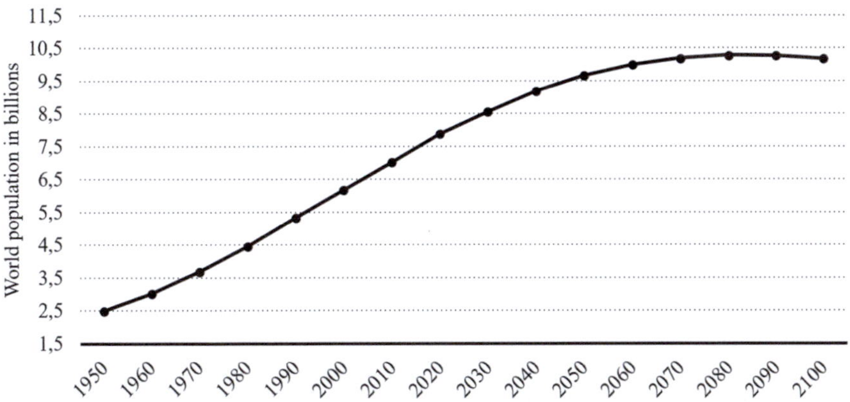

Fig. 12.3 Total population growth [123]

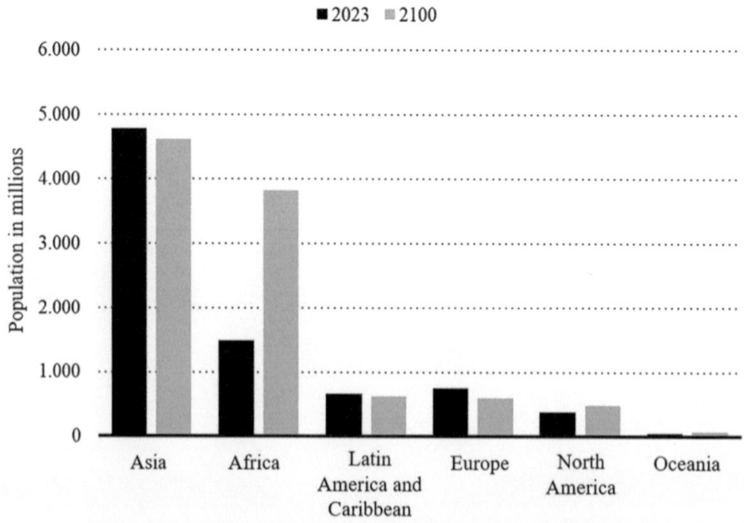

Fig. 12.4 Projected population growth by different regions of the world [124]

impacts on global ecological, economic, and social systems and, therefore, require comprehensive interdisciplinary engagement and the development of sustainable solutions.

Another relevant statistic selected for the analysis of this complex of topics is the breakdown of population growth by different regions or continents. The following statistic illustrates in which geographical areas of the world the population growth is projected to be by 2100 (Fig. 12.4).

A comprehensive analysis of the social consequences of population growth is not carried out here. However, it can be noted that this growth brings significant challenges and potential tensions for global society. These require a targeted response through appropriate measures to minimise negative impacts and promote sustainable development goals.

12.2 Ecology

The second section of this chapter focuses on the field of ecology. Following the already established approach, an overview of the current situation and relevant discussion aspects is created using selected and considered relevant statistical data.

A first aspect in the context of ecology will be introduced and discussed through the concept of the *Earth Overshoot Day*. This day is determined annually as part of a campaign by the *Global Footprint Network*. It marks the day in the respective year when human consumption of naturally available resources exceeds the supply and the Earth's ability to regenerate these resources in the same year. For the

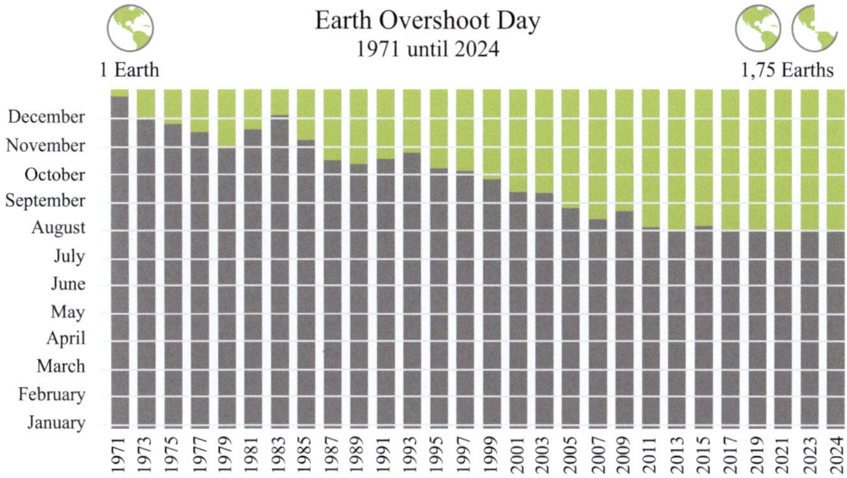

Fig. 12.5 Earth overshoot day [125]

discussion and evaluation of this concept, an overview of the development from 1971 to 2024 is selected and graphically represented in the following Figure 12.5.

Upon examination, a clear trend emerges. As can be seen, the Earth Overshoot Day occurs earlier each year. Conversely, this means that the Earth's natural resources are depleted earlier each year.

In addition to the explanations on the Earth Overshoot Day, an additional figure in the context of resource overuse is presented below. Figure 12.6 lists relevant areas or resources that are particularly affected by overuse.

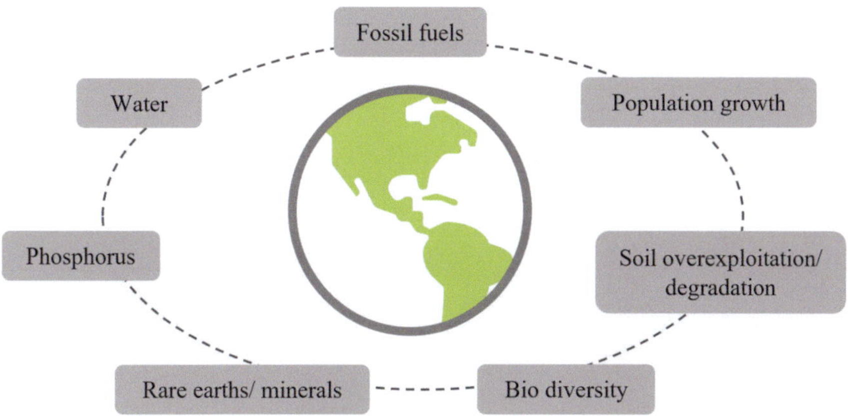

Fig. 12.6 Affected areas in the context of resource overuse; own representation based on Vajna [126]

The representation serves not only to illustrate the corresponding resources but also to convey a more detailed understanding of the complex challenges associated with global resource overuse.

Another statistic selected for this section describes the CO_2 emissions in various countries from 1990 to 2022. This representation provides a detailed overview of the development of greenhouse gas emissions in the respective states and allows for the analysis of trends and differences over time.

The figure 12.7 the development of CO_2 emissions from selected countries from 1990 to 2022. These emissions are crucial drivers of global warming, as they contribute to the increase in greenhouse gas concentrations in the Earth's atmosphere. The presented data trend allows an analysis of changes and trends in emission levels, which are essential for understanding the sectoral and geographical differences in contributions to global warming. It can be seen that a limited number of countries cause a large proportion of CO_2 emissions. This concentration of emissions in certain geographical regions provides a basis for understanding how different economic and industrial structures contribute to global warming. From an ecological perspective, this distribution underscores the need to recognise the specific contribution and responsibility in the global context. Insights of this kind are crucial in formulating targeted measures and strategies to promote sustainable development aimed at reducing emissions and preserving the environment.

A final addition to the topic of ecology is the business-as-usual scenario of the World 3 model shown in Figure 12.8.

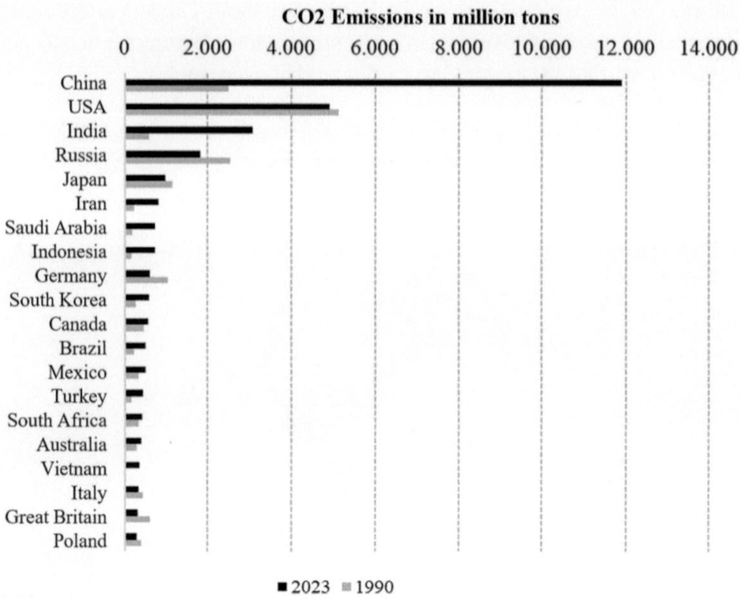

Fig. 12.7 CO_2 emissions by countries compared between 1990 and 2023 [127]

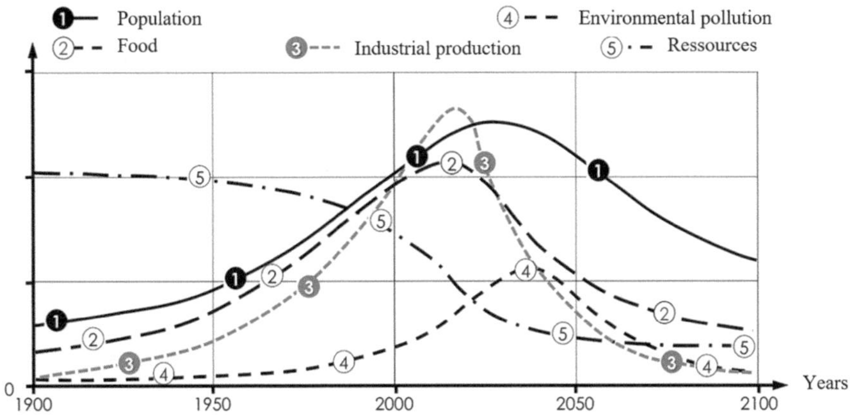

Fig. 12.8 Business-as-usual scenario of the World 3 model according to Meadows [119], figure after Vajna [126]

The model complements and details the development paths of population, food, industrial production, environmental pollution, and resources already introduced in Figure 12.1, in the event that no changes in current practices occur, and also shows potential trajectories and necessary actions in the context of ecology.

In summary of the data and facts presented, it becomes clear that the global society in 2025 is confronted with both an increasing scarcity of resources and an increased population orientation. This dynamic poses significant challenges, as the rising demand for limited resources in conjunction with population growth can potentially lead to increased social and economic tensions. The developments outlined paint an increasingly negative trend scenario, pointing to the need for rapid, efficient, and effective action. This could be supported in the long term through the implementation of sustainable development strategies, the promotion of technological innovations for resource efficiency, and increased international cooperation.

12.3 Economics

As initially described, the final section of this chapter will focus on observations and developments in the field of economics. In line with the previous sections, a picture of the current, predominantly economic situation will be presented using selected and relevant backgrounds, to subsequently derive resulting scenarios and create an understanding of which action fields need to be focused on in the long term.

A first statistic considered relevant for the topic of economics is the Gross Domestic Product (short: GDP) of countries. In the following Figure 12.9, this is

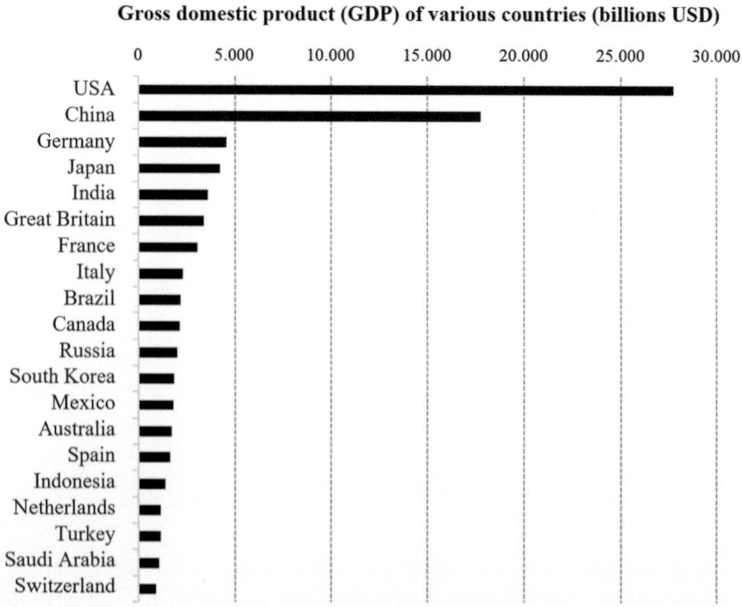

Fig. 12.9 Gross domestic product in billions of US dollars by countries in 2023 [128]

broken down by the 20 countries with the highest values in 2023 in billions of US dollars.

The analysis of the statistics shows how significantly the Gross Domestic Products of the world differ. These differences illustrate the economic inequalities between countries. Furthermore, there is a direct correlation between GDP and the burden on natural resources, as higher economic performance often goes hand in hand with more intensive resource use. These insights underline the importance of a balanced development policy to promote both economic growth and sustainable resource usage.

After the urgent need for acute and efficient action has been outlined in the preceding sections, the question arises as to how such an endeavour can be initiated. Especially in the context of economics, which represents the economic basic system according to which societies align themselves, various approaches for fundamental changes can be identified. In the following Table 12.1, alternative economic and societal concepts according to Vajna [124] are compiled. These concepts represent various attempts or solution approaches to present novel economic concepts with which societal change and a sustainability-oriented development could be initiated.

Table 12.1 Alternative economic concepts according to Vajna [126]

	Goal/vision	Representative	Perspective	Approach
Green economy	Green restructuring of the economy leads to sustainable development	UNEP (UN Environment Programme), OECD (Organisation for Economic Co-operation and Development)	National and international	Political
Europe 2020	Decoupling is possible through intelligent, sustainable, and inclusive growth	European Commission, European Council	European Union	Political
Enquiry commission for growth, prosperity and quality of life	Concrete political recommendations create more prosperity and quality of life in Germany	17 Members of the German Bundestag (legislative period 2017-2021); 17 external experts	Germany in an international context	Parliamentary debate
Cradle to cradle	Closed material cycles enable "intelligent waste"	Michael Braungart, William McDonough	International	Scientific, entrepreneurial
Steady state economy	Economic development at an optimal physical level	Herman E. Daly	Global	Scientific

Bibliography

1. Gerlach LP (1991) Global Thinking, Local Acting. Eval Rev 15:120–148. https://doi.org/10.1177/0193841X9101500107
2. King A (1991) The first global revolution: A report by the Council of the Club of Rome, 1. ed. Current affairs. Pantheon Books, New York
3. Hauff M von (2014) Nachhaltige Entwicklung. OLDENBOURG WISSENSCHAFTSVERLAG
4. Engagement Global gGmbH. Service für Entwicklungsinitiativen (2024) 17 Ziele für Nachhaltige Entwicklung. https://17ziele.de/downloads.html. Accessed 16 Sep 2024
5. Die Bundesregierung (2023) Empfehlungen zur Einbeziehung von Nachhaltigkeitszielen bei der Gesetzgebung. Staatssekretärsausschuss für nachhaltige Entwicklung, Berlin
6. Hubka V (1976) Theorie der Konstruktionsprozesse. Springer Berlin Heidelberg, Berlin, Heidelberg
7. Schaltegger S, Petersen H (2009) Corporate Social Responsibility (CSR) nachhaltig im Unternehmen verankern. Eine Herausforderung an die Managementbildung. JSSE - Journal of Social Science Education, 3–2009: Nachhaltigkeit und fachdidaktische Herausforderungen/Sustainability. https://doi.org/10.4119/jsse-413
8. Wissenschaftlicher Beirat der Bundesregierung Globale Umweltveränderungen (2011) Welt im Wandel: Gesellschaftsvertrag für eine Große Transformation ; [Hauptgutachten, 2., veränd. Aufl. Wiss. Beirat der Bundesregierung Globale Umweltveränderungen (WBGU), Berlin
9. Sato H (2024) Ikigai: The Japanese Secret: The Guide to Living Life to the Fullest. Amazon Digital Services LLC - Kdp
10. VDI/VDE-Gesellschaft Mess- und Automatisierungstechnik (2021) VDI-Richtlinie 2206: Entwicklung mechatronischer und cyper-physischer Systeme
11. VDI-Gesellschaft Produkt- und Prozessgestaltung (2019) VDI-Richtlinie 2221: Blatt 1: Entwicklung technischer Produkte und Systeme. Modell der Produktentwicklung
12. McAloone TC, Niki Bey (2009) Environmental improvement through product development: A guide. Danish Environmental Protection Agency
13. (2023) Gesetz zum Neustart der Digitalisierung der Energiewende
14. Umweltbundesamt (2025) Der Europäische Emissionshandel. https://www.umweltbundesamt.de/daten/klima/der-europaeische-emissionshandel#vergleich-von-emissionen-und-emissionsobergrenzen-cap-im-stationaren-eu-ets-1. Accessed 09 Apr 2025
15. Teich E, Brodhun C, Claus T (2015) Einsatz der Szenariotechnik in der Produktionsplanung. In: Claus T, Herrmann F, Manitz M (eds) Produktionsplanung und –steuerung. Springer Berlin Heidelberg, Berlin, Heidelberg, pp 61–88
16. Gausemeier J, Fink A, Schlake O (1998) Scenario Management. Technological Forecasting and Social Change 59:111–130. https://doi.org/10.1016/S0040-1625(97)00166-2
17. Gausemeier J, Plass C (2014) Zukunftsorientierte Unternehmensgestaltung: Strategien, Geschäftsprozesse und IT-Systeme für die Produktion von morgen, 2., überarb. Aufl. Hanser eLibrary. Hanser, München

R. Lachmayer et al., *Methodology for the Development of Sustainable Products*,
https://doi.org/10.1007/978-3-662-72442-2

18. Habicher D, Windegger F, Gruber M et al. (2020) Denkanstoß Covid-19: Zukunftsszenarien für ein nachhaltiges Südtirol 2030+, Bozen
19. Pfeiffer W (1982) Technologie-Portfolio zum Management strategischer Zukunftsgeschäftsfelder. Innovative Unternehmensführung, vol 7. Vandenhoeck & Ruprecht, Göttingen
20. Albers S (2005) Handbuch Technologie- und Innovationsmanagement: Strategie — Umsetzung — Controlling. SpringerLink Bücher. Gabler Verlag, Wiesbaden, s.l.
21. Osterwalder A, Pigneur Y (2010) Business model generation: A handbook for visionaries, game changers, and challengers. Wiley, Hoboken, NJ
22. Dyckhoff H, Souren R (2007) Nachhaltige Unternehmensführung: Grundzüge industriellen Umweltmanagements. SpringerLink Bücher. Springer Berlin Heidelberg, Berlin, Heidelberg
23. DIN Deutsches Institut für Normung e.V. DIN EN ISO 26000:2021–04, Leitfaden zur gesellschaftlichen Verantwortung (ISO_26000:2010); Deutsche Fassung EN_ISO_26000:2020
24. Geschäftsstelle des Umweltgutachterausschusses (ed) (2020) Einstieg ins Umweltmanagement mit EMAS: Ein Leitfaden für Management und Beauftragte 2020
25. Bundesministerium für Umwelt, Naturschutz, nukleare Sicherheit und Verbraucherschutz (2025) EMAS (Eco-Management und Audit Scheme). https://www.bmuv.de/WS399. Accessed 17 Jan 2025
26. Bundesministerium für Klimaschutz, Umwelt, Energie, Mobilität, Innovation und Technologie (2025) Global Reporting Initiative (GRI). https://www.bmk.gv.at/themen/klima_umwelt/nachhaltigkeit/unternehmen/standards/gri.html. Accessed 17 Jan 2025
27. Umweltbundesamt (2013) Umweltdeklaratioin von Bauprodukten: Zweck, Aufbau und Erarbeitung einer EPD. https://www.umweltbundesamt.de/themen/wirtschaft-konsum/produkte/bauprodukte/umweltdeklaration-von-bauprodukten#der-zweck-einer-umweltdeklaration-fur-bauprodukte. Accessed 17 Jan 2025
28. Hallstedt SI (2017) Sustainability criteria and sustainability compliance index for decision support in product development. Journal of Cleaner Production 140:251–266. https://doi.org/10.1016/j.jclepro.2015.06.068
29. Bundesministerium der Justiz (1990) Gesetz über die Haftung für fehlerhafte Produkte (Produkthaftungsgesetz - ProdHaftG)
30. (2008) Verordnung (EG) Nr. 765/2008 des Europäischen Parlaments und des Rates vom 9. Juli 2008 über die Vorschriften für die Akkreditierung und Marktüberwachung im Zusammenhang mit der Vermarktung von Produkten und zur Aufhebung der Verordnung (EWG) Nr. 339/93 des Rates
31. Umweltbundesamt (2024) Elektro- und Elektronikgerätegesetz. https://www.umweltbundesamt.de/themen/abfall-ressourcen/produktverantwortung-in-der-abfallwirtschaft/elektroaltgeraete/elektro-elektronikgeraetegesetz#sinn-und-zweck-des-elektrog. Accessed 17 Jan 2025
32. Bundesministerium der Justiz (2012) Gesetz zur Förderung der Kreislaufwirtschaft und Sicherung der umweltverträglichen Bewirtschaftung von Abfällen (Kreislaufwirtschaftsgesetz - KrWG)
33. Bundesministerium der Justiz (2010) Verordnung zum Schutz vor Gefahrstoffen (Gefahrstoffverordnung - GefStoffV)
34. (2011) Richtlinie 2011/65/EU des Europäischen Parlaments und des Rates zur Beschränkung der Verwendung bestimmter gefährlicher Stoffe in Elektro- und Elektro- und Elektronikgeräten (RoHS-Richtlinie)
35. Roth K (2001) Konstruieren mit Konstruktionskatalogen. Springer Berlin Heidelberg, Berlin, Heidelberg
36. Pahl G, Beitz W, Gericke K et al. (2021) Grundlagen technischer Systeme. In: Bender B, Gericke K (eds) Pahl/Beitz Konstruktionslehre. Springer Berlin Heidelberg, Berlin, Heidelberg, pp 9–25
37. Deutscher Bundestag (1994) Bericht der Enquete-Kommission „Schutz des Menschen und der Umwelt - Bewertungskriterien und Perspektiven für umweltverträgliche Stoffkreisläufe in der Industriegesellschaft": Die Industriegesellschaft gestalten - Perspektiven für einen nachhaltigen Umgang mit Stoff- und Materialströmeen. Drucksache 12/1951

38. Herrmann C (2010) Ganzheitliches Life Cycle Management. Springer Berlin Heidelberg, Berlin, Heidelberg
39. Gräßler I, Pottebaum J (2021) Generic Product Lifecycle Model: A Holistic and Adaptable Approach for Multi-Disciplinary Product–Service Systems. Applied Sciences 11:4516. https://doi.org/10.3390/app11104516
40. Walther G (2010) Nachhaltige Wertschöpfungsnetzwerke: Überbetriebliche Planung und Steuerung von Stoffströmen entlang des Produktlebenszyklus. Zugl. Braunschweig, Techn. Univ., Habil.-Schr., 2009, 1. Aufl. Gabler Research. Gabler, Wiesbaden
41. Ehrlenspiel K, Kiewert A, Lindemann U et al. (2014) Kostengünstig Entwickeln und Konstruieren. Springer Berlin Heidelberg, Berlin, Heidelberg
42. Schmidt M (2021) Der Einsatz von Sankey-Diagrammen im Stoffstrommanagement. Beiträge der Hochschule Pforzheim, vol 124. Hochschule Pforzheim, Pforzheim
43. Kennedy ABW, Sankey HR (1898) The Thermal Efficiency of Steam Engines. Report of the Committee Appointed to the Council upon the Subject of the Definition of a Standard or Standards of Thermal Efficiency for Steam Engines: With an Introductory Note (Including Appendixes and Plate at Back of Volume). Minutes of the Proceedings of the Institution of Civil Engineers 134:278–312. https://doi.org/10.1680/imotp.1898.19100
44. Wolf A (2016) Laserscheinwerfer für Kraftfahrzeuge. Dissertation, Gottfried Wilhelm Leibniz Universität Hannover; TEWISS - Technik und Wissen GmbH
45. Umweltbundesamt (2023) Projektionsbericht 2023 für Deutschland, Dessau-Roßlau
46. Aamodt A, Nygård M (1995) Different roles and mutual dependencies of data, information, and knowledge—An AI perspective on their integration. Data & Knowledge Engineering 16:191–222. https://doi.org/10.1016/0169-023X(95)00017-M
47. Sonderforschungsbereich 653 (2005) Genetik und Intelligenz - Grundlagen für Industrie 4.0: Warum die Welt cyberphysisch wird - und wie gentelligente Systeme dazu beitragen. https://www.sfb653.uni-hannover.de/de/der-sfb. Accessed 10 Apr 2025
48. Hintemann R, Hinterholzer S, Clausen J (2020) Rechenzentren in Europa - Chancen für eine nachhaltige Digitalisierung - Teil 2, Berlin
49. Vries A de (2023) The growing energy footprint of artificial intelligence. Joule 7:2191–2194. https://doi.org/10.1016/j.joule.2023.09.004
50. (2008) Klimaänderung 2007: Synthesebericht, Berlin
51. Food and Agriculture Organization of the United Nations (2010) Greenhouse Gas Emissions from the Dairy Sector: A Life Cycle Assessment
52. Poore J, Nemecek T (2018) Reducing food's environmental impacts through producers and consumers. Science 360:987–992. https://doi.org/10.1126/science.aaq0216
53. Pastoors S, Scholz U (2018) Methoden zum Messen der Nachhaltigkeit von Produkten. In: Scholz U, Pastoors S, Becker JH et al. (eds) Praxishandbuch Nachhaltige Produktentwicklung. Springer Berlin Heidelberg, Berlin, Heidelberg, pp 23–30
54. Statistisches Bundesamt (2025) Regionalatlas Deutschland. https://regionalatlas.statistikportal.de/#. Accessed 11 Apr 2025
55. Seliger G (2002) Nachhaltige Technologien. In: Milberg J, Schuh G (eds) Erfolg in Netzwerken. Springer Berlin Heidelberg, Berlin, Heidelberg, pp 235–243
56. Lindemann U (2009) Methodische Entwicklung technischer Produkte. Springer Berlin Heidelberg, Berlin, Heidelberg
57. VDI/VDE-Gesellschaft Entwicklung Konstruktion Vertrieb (1997) VDI-Richtlinie 2225: Konstruktionsmethodik: Technisch-wirtschaftliches Konstruieren. Vereinfachte Kostenermittlung
58. VDI-Gesellschaft Produkt- und Prozessgestaltung (2018) VDI-Richtlinie 2028: Blatt 1: Bewerten in der Wertanalyse - Vorgehen und Werkzeuge
59. Feldhusen J, Grote K-H, Nagarajah A et al. (2013) Vorgehen bei einzelnen Schritten des Produktentstehungsprozesses. In: Feldhusen J, Grote K-H (eds) Pahl/Beitz Konstruktionslehre. Springer Berlin Heidelberg, Berlin, Heidelberg, pp 291–409

60. Ritthoff M, Rohn H, Liedtke C (2002) MIPS berechnen: Ressourcenproduktivität von Produkten und Dienstleistungen. Wuppertal spezial, vol 27. Wuppertal-Inst. für Klima Umwelt Energie, Wuppertal
61. Kambanou ML (2021) Life Cycle Costing: Supporting companies towards a circular economy, 1st ed. Linköping University Medical Dissertations Ser, v.2152. Linköping University Electronic Press, Linköping
62. DIN Deutsches Institut für Normung e.V. DIN EN ISO 14040:2021–02, Umweltmanagement_- Ökobilanz_- Grundsätze und Rahmenbedingungen (ISO_14040:2006_+ Amd_ 1:2020); Deutsche Fassung EN_ISO_14040:2006_+ A1:2020
63. Huijbregts MAJ, Steinmann ZJN, Elshout PMF et al. (2017) ReCiPe2016: a harmonised life cycle impact assessment method at midpoint and endpoint level. Int J Life Cycle Assess 22:138–147. https://doi.org/10.1007/s11367-016-1246-y
64. David R, Alla H (2005) Discrete, continuous, and hybrid Petri nets. Springer, Berlin
65. Wurst J, Ganter NV, Ehlers T et al. (2023) Assessment of the ecological impact of metal additive repair and refurbishment using powder bed fusion by laser beam based on a multiple case study. Journal of Cleaner Production 423:138630. https://doi.org/10.1016/j.jclepro.2023.138630
66. Ganter NV (2023) Reparatur und Modernisierung metallischer Bauteile durch pulverbettbasiertes Schmelzen mittels Laserstrahl. Dissertation, Gottfried Wilhelm Leibniz Universität Hannover; TEWISS - Technik und Wissen GmbH
67. Schaltegger S, Kubat R, Hilber C et al. (1996) Innovatives Management staatlicher Umweltpolitik. Birkhäuser Basel, Basel
68. Saling P (2022) Nachhaltigkeitsbewertungen und die Rolle von Standards und Normen. In: Schwager B (ed) CSR und Nachhaltigkeitsstandards. Springer Berlin Heidelberg, Berlin, Heidelberg, pp 275–288
69. Mann H, Singh Mann IJ, Gullaiya N (2017) The role of organizational orientation and product attributes in performance for sustainability. Procedia Computer Science 122:850–856. https://doi.org/10.1016/j.procs.2017.11.446
70. HPI Academy (2025) Was ist Design Thinking? https://hpi.de/d-school/themen/design-thinking/. Accessed 20 Mar 2025
71. Kerguenne A, Schaefer H, Taherivand A (2024) Design Thinking: Die agile Innovations-Strategie, 3. Auflage. Haufe Taschen Guide, vol 307. Haufe, Freiburg
72. Heike da Silva Cardoso, Nicolas Kusser, Jana Kieselstein (2024) Einsatz von Künstlicher Intelligenz bei der wissenschaftlichen Literaturrecherche: ein Überblick
73. VDI-Gesellschaft Technologies of Life Sciences (2021) VDI-Richtlinie 6220: Blatt 1: Bionik. Grundlagen, Konzeption und Strategie
74. Popov VV, Muller-Kamskii G, Kovalevsky A et al. (2018) Design and 3D-printing of titanium bone implants: brief review of approach and clinical cases. Biomed Eng Lett 8:337–344. https://doi.org/10.1007/s13534-018-0080-5
75. The Biomimicry Institute (2025) Ask Nature. https://asknature.org/. Accessed 10 Feb 2025
76. Zwicky F (1989) Entdecken, Erfinden, Forschen im morphologischen Weltbild, 2. Aufl., Repr. Baeschlin, Glarus
77. Roth K (2000) Konstruieren mit Konstruktionskatalogen. Springer Berlin Heidelberg, Berlin, Heidelberg
78. WGSN Limited, OC&C Strategy Consultants (2023) Doing more with less: Forecasting for success
79. Berg A, Granskog A, Lee L et al. Fashion On Climate: How the Fashion Industry can urgently act to reduce its greenhouse gas emissions, 2020
80. Jung C (2006) Anforderungsklärung in interdisziplinärer Entwicklungsumgebung. Dissertation
81. Heine H (2023) A high-flux cold atom source based on a nano-structured atom chip. Hannover : Institutionelles Repositorium der Leibniz Universität Hannover

82. Krause D, Vietor T, Inkermann D et al. (2021) Produktarchitektur. In: Bender B, Gericke K (eds) Pahl/Beitz Konstruktionslehre. Springer Berlin Heidelberg, Berlin, Heidelberg, pp 335–393
83. Krause D, Gebhardt N (eds) (2018) Methodische Entwicklung modularer Produktfamilien. Springer Berlin Heidelberg, Berlin, Heidelberg
84. Koppenhagen F (2014) Modulare Produktarchitekturen – Komplexitätsmanagement in der frühen Phase der Produktentwicklung. In: Schoeneberg K-P (ed) Komplexitätsmanagement in Unternehmen. Springer Fachmedien Wiesbaden, Wiesbaden, pp 113–162
85. Rathnow PJ (1993) Integriertes Variantenmanagement: Bestimmung, Realisierung und Sicherung der optimalen Produktvielfalt. Innovative Unternehmensführung, Bd. 20. Vandenhoeck & Ruprecht, Göttingen
86. Potting J, Hekkert MP, Worrell E et al. (2017) Circular Economy: Measuring innovation in the product chain
87. Kirchherr J, Reike D, Hekkert M (2017) Conceptualizing the Circular Economy: An Analysis of 114 Definitions. SSRN Journal. https://doi.org/10.2139/ssrn.3037579
88. Spektrum der Wissenschaft (1998) Peltier-Effekt. https://www.spektrum.de/lexikon/physik/peltier-effekt/11003. Accessed 26 Feb 2025
89. Ponn J, Lindemann U (2011) Konzeptentwicklung und Gestaltung technischer Produkte. Springer Berlin Heidelberg, Berlin, Heidelberg
90. Siefer T, Schütze C (2019) Öffentlicher Nahverkehr, Eisenbahnbau und Bahnbetrieb. In: Zilch K, Diederichs CJ, Beckmann KJ et al. (eds) Handbuch für Bauingenieure. Springer Fachmedien Wiesbaden, Wiesbaden, pp 1–46
91. Eberle RF (1972) Developing Imagination Through Scamper*. Journal of Creative Behavior 6:199–203. https://doi.org/10.1002/j.2162-6057.1972.tb00929.x
92. Albers A, Bursac N, Wintergerst E Produktgenerationsentwicklung - Bedeutung und Herausforderungen aus einer entwicklungsmethodischen Perspektive. In: Stuttgarter Symposium für Produktentwicklung (SSP) : Stuttgart, 19. Juni 2015 ; Hrsg.: H. Binz
93. Ehrlenspiel K (2009) Integrierte Produktentwicklung: Denkabläufe, Methodeneinsatz, Zusammenarbeit, 4., aktualisierte Aufl. Hanser, München, Wien
94. Pahl G, Beitz W (2021) Gestaltungsprinzipien. In: Bender B, Gericke K (eds) Pahl/Beitz Konstruktionslehre. Springer Berlin Heidelberg, Berlin, Heidelberg, pp 525–565
95. Spura C, Fleischer B, Wittel H et al. (2023) Kupplungen und Bremsen. In: Spura C, Fleischer B, Wittel H et al. (eds) Roloff/Matek Maschinenelemente. Springer Fachmedien Wiesbaden, Wiesbaden, pp 461–537
96. Bender B, Gericke K (2021) Pahl/Beitz Konstruktionslehre. Springer Berlin Heidelberg, Berlin, Heidelberg
97. Bundesverband WindEnergie e.V. Schallemissionen vonWindenergieanlagen im Verhältnis zu anderen Lärmquellen. https://www.wind-energie.de/fileadmin/redaktion/dokumente/publikationen-oeffentlich/Service/Infografiken/20181010_Vergleich_Schallemmissionen.pdf. Accessed 04 Apr 2025
98. Charter M, Tischner U (2017) Sustainable Solutions. Routledge
99. McAloone TC, Pigosso DCA (2021) Ökodesign. In: Bender B, Gericke K (eds) Pahl/Beitz Konstruktionslehre. Springer Berlin Heidelberg, Berlin, Heidelberg, pp 975–1021
100. Luttropp C, Lagerstedt J (2006) EcoDesign and The Ten Golden Rules: generic advice for merging environmental aspects into product development. Journal of Cleaner Production 14:1396–1408. https://doi.org/10.1016/j.jclepro.2005.11.022
101. Lachmayer R, Ehlers T, Lippert RB (2022) Entwicklungsmethodik für die Additive Fertigung, 2. Auflage. Lehrbuch. Springer Vieweg, Berlin, Heidelberg
102. Pahl G, Beitz† W, Feldhusen J et al. (2013) Gestaltungsrichtlinien. In: Feldhusen J, Grote K-H (eds) Pahl/Beitz Konstruktionslehre. Springer Berlin Heidelberg, Berlin, Heidelberg, pp 583–751
103. DIN Deutsches Institut für Normung e.V. DIN EN 16575:2014–10, Biobasierte Produkte_-Terminologie; Deutsche Fassung EN_16575:2014

104. Saulich K (2016) Ressourceneffizienz biobasierter Materialien im verarbeitenden Gewerbe, Berlin
105. Fuqua MA, Huo S, Ulven CA (2012) Natural Fiber Reinforced Composites. Polymer Reviews 52:259–320. https://doi.org/10.1080/15583724.2012.705409
106. Schürmann H (2007) Konstruieren mit Faser-Kunststoff-Verbunden: Mit 39 Tabellen, 2., bearb. und erw. Aufl. VDI-/Buch]. Springer, Berlin, Heidelberg
107. Vogt D, Karus M, Ortmann S et al. (2006) Wood-Plastic-Composites (WPC) Holz-Kunststoff-Verbundwerkstoffe: Märkte in Nordamerika, Japan und Europa mit Schwerpunkt auf Deutschland Technische Eigenschaften – Anwendungsgebiete Preise – Märkte – Akteure, Hürth
108. Fachagentur Nachwachsende Rohostoffe e.V. (2020) Nachwachsende Materialien für leichte Konstruktionen: Neue technische Bio-Verbundwerkstoffe für Automobil- und Elektroindustrie. https://news.fnr.de//fnr-pressemitteilung/nachwachsende-materialien-fuer-leichte-konstruktionen. Accessed 13 Mar 2025
109. Obata Y, Takeuchi K, Furuta Y et al. (2005) Research on better use of wood for sustainable development: Quantitative evaluation of good tactile warmth of wood. Energy 30:1317–1328. https://doi.org/10.1016/j.energy.2004.02.001
110. Bundesministerium für Wirtschaft und Klimaschutz (2025) Holz im Maschinenbau. Industrielle Bioökonomie in Deutschland. https://www.bmwk.de/Redaktion/DE/Dossier/Industrielle-Biooekonomie/Best-Practice-Beispiele/026-ligenium/00-best-practice-beispiel.html. Accessed 28 Feb 2025
111. Engelhardt M, Khaloian Sarnaghi A, Buchelt B et al. (2022) Holzbasierte Werkstoffe im Maschinenbau (HoMaba) - Berechnungskonzepte, Kennwertanforderungen, Kennwertermittlung : Schlussbericht zum Vorhaben : Laufzeit: 01.11.2018 bis 30.04.2022. Technische Universität München (TUM)
112. Knott M (2023) Viele Vorteile: Das ist der weltweit größte Windrad-Turm aus Holz. https://www.netzwelt.de/news/219189-viele-vorteile-weltweit-groesste-windrad-turm-holz.html. Accessed 12 May 2025
113. Poll D (2024) Leicher und nachhaltiger: Holz als Trumpf im Maschinenbau. https://www.produktion.de/technik/co2-neutrale-industrie/leichter-und-nachhaltiger-holz-als-trumpf-im-maschinenbau-675.html. Accessed 28.02.20258
114. Vajna S (ed) (2014) Integrated Design Engineering: Ein interdisziplinäres Modell für die ganzheitliche Produktentwicklung. Springer-Vieweg, Berlin, Heidelberg
115. Biedermann H, Kinz A (2021) Lean Smart Maintenance: Agiles, lern- und wertschöpfungsorientiertes Instandhaltungsmanagement. Research. Springer Gabler, Wiesbaden, Heidelberg
116. DIN Deutsches Institut für Normung e.V. DIN 31051:2019–06, Grundlagen der Instandhaltung
117. DIN Deutsches Institut für Normung e.V. DIN EN ISO 9001:2015–11, Qualitätsmanagementsysteme_- Anforderungen (ISO_9001:2015); Deutsche und Englische Fassung EN_ISO_9001:2015
118. Wissenschaftliche Gesellschaft für Produktentwicklung WiGeP e.V (2022) Update-Factory für ein industrielles Produkt-Update: Ein Beitrag zur Kreislaufwirtschaft. Positionspapier Update-Factory 2022
119. Meadows DH (2005) The limits to growth: The 30-year update, Rev. ed. Earthscan, London
120. Herrington G (2021) Update to limits to growth: Comparing the World3 model with empirical data. J of Industrial Ecology 25:614–626. https://doi.org/10.1111/jiec.13084
121. Corsten H, Roth S (eds) (2012) Nachhaltigkeit. Gabler Verlag, Wiesbaden
122. Johan Rockström, Pavan Sukhdev (2016) The SDGs wedding cake. https://www.stockholmresilience.org/research/research-news/2016-06-14-how-food-connects-all-the-sdgs.html. Accessed 10 Jan 2025
123. United Nations Department of Economic and Social Affairs (UN DESA) (2024) Prognose zur Entwicklung der Weltbevölkerung von 1950 und 2100 (in Milliarden). In Statista. https://de.statista.com/statistik/daten/studie/1717/umfrage/prognose-zur-entwicklung-der-weltbevoelkerung/. Accessed 15 Apr 2025

124. United Nations Department of Economic and Social Affairs (UN DESA) (2024) Prognostiziertes Bevölkerungswachstum nach unterschiedlichen Regionen der Welt. In Statista. https://de.statista.com/statistik/daten/studie/184686/umfrage/weltbevoelkerung-nach-kontinenten/. Accessed 15 Apr 2025
125. Global Footprint Network (2024) Earth Overshoot Day 2024. https://overshoot.footprintnetwork.org/. Accessed 10 Jan 2025
126. Vajna S (ed) (2022) Integrated Design Engineering. Springer Berlin Heidelberg, Berlin, Heidelberg
127. Global Carbon Project (2024) Höhe der CO2-Emissionen nach ausgewählten Ländern weltweit im Jahresvergleich 1990 und 2023. In Statista. https://de.statista.com/statistik/daten/studie/167864/umfrage/co-emissionen-in-ausgewaehlten-laendern-weltweit/
128. International Monetary Fund (2023) Ranking de 20 Länder mit dem größten Bruttoinlandsprodukt (BIP) im Jahr 2023 (in Milliarden US-Dollar). In Statista. https://de.statista.com/statistik/daten/studie/157841/umfrage/ranking-der-20-laender-mit-dem-groessten-bruttoinlandsprodukt/

MIX
Papier aus verantwortungsvollen Quellen
Paper from responsible sources
FSC® C105338

If you have any concerns about our products,
you can contact us on
ProductSafety@springernature.com

In case Publisher is established outside the EU,
the EU authorized representative is:
Springer Nature Customer Service Center GmbH
Europaplatz 3, 69115 Heidelberg, Germany

Printed by Libri Plureos GmbH
in Hamburg, Germany